Estrogens - Recent Advances

Edited by Courtney Marsh

Published in London, United Kingdom

Estrogens - Recent Advances
http://dx.doi.org/10.5772/intechopen.100987
Edited by Courtney Marsh

Contributors
Pitu Wulandari, Luz Irene Pascual Mathey, José Locia-Espinoza, Shigeru Takami, Sawa Horie, Hong-Bing Zhang, Yao Wang, Bing Wang, Courtney Marsh

First published in London, United Kingdom, 2023 by IntechOpen
IntechOpen is the global imprint of INTECHOPEN LIMITED, registered in England and Wales, registration number: 11086078, 5 Princes Gate Court, London, SW7 2QJ, United Kingdom

British Library Cataloguing-in-Publication Data
A catalogue record for this book is available from the British Library

Additional hard and PDF copies can be obtained from orders@intechopen.com

Estrogens - Recent Advances
Edited by Courtney Marsh
p. cm.
Print ISBN 978-1-80356-200-1
Online ISBN 978-1-80356-201-8
eBook (PDF) ISBN 978-1-80356-202-5

Meet the editor

Dr. Courtney Marsh is an associate professor in the Departments of Obstetrics and Gynecology and Cell Biology and Physiology, at the University of Kansas School of Medicine (KUMC). She is the KUMC Reproductive Endocrinology and Infertility Division Director, joining the practice in 2013. Dr. Marsh is actively involved with hormonal research involving estrogen and has multiple publications in sex steroid feedback and hormonal therapy to her credit. She enjoys tennis and spending time with her family.

Contents

Preface

Estrogens - Recent Advances provides updated information on the sex steroid hormone estrogen and the many roles it plays in the body. Most readers may be familiar with the function of estrogen in the hypothalamic–pituitary–ovarian/uterine axis. With this book, we move beyond the traditional role of estrogen action on the ovaries and uterus and discuss exciting areas of new research.

Chapter 1 "Introductory Chapter: Estrogen – Sex Steroid Hormone in Ovary and beyond" by Dr. Courtney Marsh reviews the many roles of estrogen in the human body.

In Chapter 2, "The Research Advances in G-Protein-Coupled Estrogen Receptor", Dr. Zhang Hongbing et al. explore the role of G proteins and the G-protein estrogen receptor in the human body. The author reviews the role of G-protein estrogen receptor agonists and antagonists in the context of potential modalities for patient treatment. The author also discusses various signaling pathways, including nuclear factor kappa B, Hippo, mitogen-activated protein kinase, phosphatidylinositol 3-kinase, and extracellular signal-regulated kinase, as they relate to human disease processes. This chapter highlights the key role between understanding receptor signaling and disease process to better understand the role of estrogen in the human body.

In Chapter 3, "Effect of Hypoestrogenism on Oral Cavity", Dr. Pitu Wulandari explores the effect of estrogen on the oral cavity. The author reviews a variety of oral disease states and discusses the effect of declining estrogen with menopause on oral health. In Chapter 4, "Functional Roles of Estradiol in the Olfactory and Vomeronasal Mucosae of Mammals: A Working Hypothesis", Shigeru Takami and Sawa Horie discuss the role of estrogen in olfactory mucosae and uses animal models to explore the novel topic of estrogen signaling as a potential model for human health. Both chapters look beyond the traditional role of estrogen in the ovary and uterus and focus on how estrogen affects oral and olfactory health.

Finally, in Chapter 5, "Permissive Role of Estrogens in Prostate Diseases", José Locia Espinoza and Luz Irene Pascual Mathey discuss the role of estrogen in the prostate. He reviews estrogen signaling in the prostate as well as disease states that are modified by estrogen.

Overall, this book provides the reader with a better understanding of the variety of effects estrogen has on the body. Thank you to all the authors and to the staff at IntechOpen for making this highly informative, novel book!

Courtney Marsh MD, MPH
Associate Professor Department OBGYN,
Associate Professor Cell Biology and Physiology,
University of Kansas School of Medicine,
Kansas City, KS, USA

Section 1

Estrogen Signaling

Chapter 1

Introductory Chapter: Estrogen – Sex Steroid Hormone in Ovary and beyond

Courtney Marsh

1. Introduction

1.1 Non-reproductive health effects of sex steroid hormones

Estrogens are members of the estrane family of sex steroid hormones essential to reproductive and ovarian function. Estranes are derived from cholesterol (specifically low-density lipoprotein) and undergo a series of enzymatic modifications to reduce the number of carbons contained in the molecule from 27 (cholestanes) to 18 (estranes) [1]. The movement of cholesterol from the mitochondrial outer membrane of the cell to the inner membrane is the rate-limiting step for steroidogenesis and estrogen production.

Estrone, estradiol, estriol, and estetrol are all estrogens in the estrane family and vary according to molecular structure and biological activity. Estrone (E1) has only one hydroxyl group and is mainly produced by peripheral aromatization from androgens. Of the estranes, estrone is predominant in postmenopausal women. In terms of bioactivity, estrone is a weak estrane. Estradiol (E2) has two hydroxyl groups and is the main estrane found in reproductive age persons with ovaries. Estradiol is also produced by the testes, liver, adrenal glands, breasts, and placenta. Estradiol has high bioactivity and rapidly declines with ovarian aging and menopause. Estriol (E3) has three hydroxyl groups and is the main estrane of pregnancy. Estriol is derived from androstendione, estrone, and estradiol. Although estriol has a lower affinity to the sex hormone binding globulin and higher bioavailability, it is a weaker estrange than estradiol [2]. With four hydroxyl groups, estetrol is produced solely through the fetal liver by 15 alpha and 16 alpha hydroxylation. Estetrol is a metabolite of estriol and estradiol and is found in serum and urine with peak levels at the end of pregnancy [3].

For reproductive function, estradiol is mediated through the hypothalamic–pituitary-gonadal axis. Gonadotropin-releasing hormone is released in a pulsatile fashion from the hypothalamus to stimulate follicle-stimulating hormone (FSH) from the anterior pituitary. FSH then stimulates gonads for folliculogenesis in the ovary and spermatic production in the testes.

Estrogen action is mediated through estrogen receptor signaling. Estrogen action can occur through genomic effect (estrogen receptor complex binding to DNA promoter) and non-genomic effect (no direct binding of estrogen receptor complex to DNA) [4]. Various signaling pathways are involved with the G-protein estrogen receptor including nuclear factor-kappa B, hippo, mitogen-activated protein kinase,

IntechOpen

phosphatidylinositol 3-kinase, and extracellular regulated kinase. These pathways are involved in many disease processes outside of the reproductive tract.

Estrogens – Recent Advances provides an in-depth look at the role of estrogen and the multiple organ systems it affects. As most are familiar with the role of estrogen in the ovary and uterus, this book moves beyond these organs to explore estrogen's effect on cancers, prostate, oral mucosa, and olfactory mucosa to provide an excellent review of the sex steroid hormone. By understanding the extensive role of estrogen in multiple organ systems, we may uncover novel management and treatment options for individuals in the future.

Author details

Courtney Marsh
Division of Reproductive Endocrinology and Infertility, University of Kansas School of Medicine, Kansas, USA

*Address all correspondence to: cmarsh2@kumc.edu

References

[1] Kumar A, Banerjee A, Singh D, Thakur G, Kasarpalkar N, Gavali S, et al. Estradiol: A steroid with multiple facets. Hormone and Metabolic Research. 2018;**50**(5):359-374

[2] Ali ES, Mangold C, Peiris AN. Estriol: Emerging clinical benefits. Menopause. 2017;**24**(9):1081-1085

[3] Holinka CF, Diczfalusy E, Coelingh Bennink HJT. Estetrol: A unique steroid in human pregnancy. The Journal of Steroid Biochemistry and Molecular Biology. 2008;**110**(1-2):138-143

[4] Fuentes N, Silveyra P. Estrogen receptor signaling mechanisms. Advances in Protein Chemistry and Structural Biology. 2019;**116**:135-170

Chapter 2

The Research Advances in G-Protein-Coupled Estrogen Receptor

Hong-Bing Zhang, Yao Wang and Bing Wang

Abstract

Estrogen binds to the typical estrogen receptor (ER) ERα or ERβ and is translocated to the nucleus, where it binds directly to the estrogen response element of the target gene to induce transcription and regulate gene expression, and the whole process is completed in several hours to several days. The G protein-coupled estrogen receptor (GPER), a type that is structurally distinct from typical ERα and ERβ, rapidly induces most non-genomic effects within seconds to minutes. GPER regulates cell growth, migration, and programmed cell death in a variety of tissues and has been associated with the progression of estrogen-associated cancers. Here, the characteristics, cell signal transduction, and the latest research progress of GPER in estrogen-associated tumors and retinal diseases are reviewed.

Keywords: estrogen receptor, G protein-coupled estrogen receptor, G-protein coupled receptor 30, GPER, GPR30

1. Introduction

Estrogen, the main sex hormone in cisgender women, plays an important role in regulating various physiological and pathological processes in the body and affects mood, bone strength, and even heart health beyond fertility and sex-related functions. Typical estrogen receptor (ER) includes ERα and ERβ. After combined with ERα or ERβ, estrogen transfers in nucleus and directly binds to the estrogen response element in the promoter region of target gene and then induces transcription and regulates gene expression, and this process is called estrogen genomic effect, which usually takes several hours to several days to complete this process [1].

Besides, by combined with a novel type of ER, G protein-coupled ER (GPER), an alternate ER with a structure distinct from the two canonical ERα and ERβ, can also induce rapid non-genomic effects in seconds to minutes and mediate most of the non-genomic effects of estrogen.

GPER triggers several downstream pathways that exert vital biological roles in the regulation of cell growth, migration, and programmed cell death in a variety of tissues. In particular, there is a significant correlation between GPER and the progression of estrogen-related cancers. Therefore, better understanding the role of GPER in physiological function, especially in estrogen-related cancers, may help identify new biological regulatory targets, as well as cancer therapeutic targets. In this review, we summarize the characteristic and cell signaling and detail the functions of GPER, as well as the latest research progress of GPER in estrogen-related cancers.

2. Brief overview of GPER

2.1 G-protein and GPCR

G proteins, also called as guanosine nucleotide-binding proteins, are proteins that are involved in delivering a variety of stimuli from extracellular environment to intracellular organelle. G protein-coupled receptors (GPCRs) are a general name of membrane protein receptors, which locate on cell membranes and are the largest family of proteins encoded by the human genome [2]. The unique feature of this kind of GPCRs is that there are seven transmembrane α helices in the stereo-structure of the receptor, and there are G protein-binding sites on the C-terminal of the peptide chain and the intracellular ring (the third intracellular ring) connecting the fifth and sixth transmembrane helices (starting from the N-terminal of the peptide chain). Endogenous ligands of GPCR include many compounds, such as amines, carbohydrates, lipids, peptides, proteins, hormones, neurotransmitters, chemokines, and even photons and odors. GPCR has been linked to type 2 diabetes, obesity, depression, cancer, Alzheimer's disease, and other diseases [3].

2.2 ER

Estrogen receptor (ER) is a steroid receptor, belonging to a superfamily of proteins, whose function is to regulate transcription of gene pools in other cells. The estrogen receptors such as ERα, ERβ, and GPER are located in the tissues of the female reproductive tract and breast, as well as in a variety of tissues such as bone, brain, liver, colon, skin, and salivary glands, exerting the effects of estrogen compounds on their target tissues [4].

2.3 GPER

In 1996, the third estrogen receptor, G-protein coupled receptor 30 (GPR30), was discovered in breast cancer tissues for the first time [5, 6]. GPR30 is a member of GPCR family, which can be exclusively bound to by estrogen and estrogen-like molecules, and it is involved in rapid non-genomic estrogen effects [7–9]. GPR30 also has another name GPER or GPER1 [10]. The function of this receptor provides an important basis for expounding the rapid response of estrogen. The human GPER gene, including a 1128 bp open reading frame that encodes 375-amino acid receptors, is located on chromosome 7 [6, 11].

GPER is expressed in various tissues, such as nerve, reproductive, digestive, and muscle organs [12], as well as in brain pericellular, cortex, dentate gyrus, anterior pituitary, and adrenal medulla cells [13]. The expression of GPER is not limited to the cell surface, and the binding domain of this receptor is also expressed in the endoplasmic reticulum [14].

In normal tissue, GPER was expressed markedly in the cortex and the anterior pituitary, islets and pancreatic ducts, fundic glands of the stomach, the epithelium of the duodenum and gallbladder, hepatocytes, proximal tubules of the kidney, the adrenal medulla, and syncytiotrophoblasts and decidua cells of the placenta [15].

2.4 GPER agonists and antagonists

There are synthetic and natural estrogen compounds that can interact with GPER. GPER agonists can be categorized as follows: (1) natural estrogens including 17β-estradiol (E2, a general ER agonist) and 2-methoxy-estradiol; (2) phytoestrogens including genistein, quercetin, equol, and resveratrol; (3) selective estrogen receptor modulators (SERMs), which includes tamoxifen (TMX), 4-hydroxytamoxifen, and raloxifene; (4) selective estrogen receptor downregulation agents (SERDs), which include fulvestrant and pesticides; (5) plastic compounds as endocrine disruptors, which include atrazine, bisphenol A, zearalenone, nonylphenol, and ketone. GPER antagonists can be categorized as follows: (1) natural estrogens, which include 2-hydroxyoestradiol and estriol; and (2) synthesis of compounds such as MIBE [15–17].

In addition, there are highly selective GPER ligands, which include GPER agonist G-1 and GPER antagonists G-15 and G-36 [18, 19]. G-1, also known as Tespria, is a nonsteroidal compound and a selective GPER1 agonist. *In vitro* cell experiments have shown that G-1 has no binding affinity for ERα and ERβ [20, 21]. Moreover, G-1 regulation is involved in intracellular calcium [Ca2+] and phosphoadenosine 3-kinases (PI3Ks) signaling pathways. G-15 is a selective GPER1 antagonist with high binding affinity to GPER1 and less binding to ERα and ERβ [19, 22, 23]. As another selective GPER1 antagonist, G36 also owns poor binding ability to ERα and ERβ as well as G15, whereas G36 is more powerful than G15 in binding to GPER1 and then modulates Ca2+ mobilization and PI3K signaling [18].

3. GPER-related signaling pathway

3.1 Nuclear factor-Kappa B (NF-κB) signaling pathway

When epithelial cells undergo epithelial-mesenchymal transition (EMT), mastitis may be aggravated. GPER1 activation inhibits EMT of goat mammary epithelial cells through NF-κB signaling pathway, thus impeding the occurrence of mastitis [24]. Okamoto et al. [25] delineated that lipopolysaccharide (LPS)-induced interleukin-6 (IL-6) overexpression in mouse macrophages is negatively regulated by GPER agonist G-1 *via* NF-κB signaling pathway, which is associated with anti-inflammatory effects of estrogen.

3.2 Hippo signaling pathway

Zhou et al. [26] determined that, compared with adjacent normal tissue, the expression of GPER is highly upregulated in the cancer cells of breast invasive ductal carcinoma, and key downstream signals of GPER are associated with the Hippo/YAP (Yes-associated protein 1)/TAZ (transcriptional coactivator with a PDZ-binding domain) pathway, which plays an important role in breast tumorigenesis. YAP is a pivotal effector of Hippo pathway, and Deng et al. [27] exhibited that bisphenol S activates YAP, upgrades YAP nuclear accumulation, and regulates its downstream genes in triple-negative breast cancer cells. By inhibiting GPER, the effect of bisphenol S-induced Hippo/YAP signal pathway can be weakened.

3.3 Mitogen-activated protein kinase (MAPK) signaling pathway

Triclosan can act on GPER to induce estrogen effect. It has been found that triclosan can upregulate the expression of mir-144 by activating GPER, which leads to abnormal regulation of neuro-related genes and neurodevelopmental toxicity, and this process is closely related to the activation of downstream protein kinase C (PKC)/MAPK signaling pathway [28]. In ER-positive breast cancer cells, the activation of GPER and its downstream MAPK/ERK signal pathway increase the expression level of TRIM2 protein, which leads to the reduction of Bim protein in tamoxifen-resistant breast cancer cells, providing a new idea for solving tamoxifen resistance in ER-positive breast cancer cells [29].

3.4 Phosphatidylinositol 3-kinase (PI3K)/Akt signaling pathway

The nuclear ER is negative in breast cancer SKBR-3 cells, while GPER is positive. Using the SKBR-3 cells, Shi et al. [30] demonstrated that the expression of PI3K and p-AKT can be downregulated by cryptotanshinone (an active compound of traditional Chinese medicine Danshen), and the inhibitory effect of cryptotanshinone on SKBR-3 cells is carried out by inhibiting GPER-mediated PI3K/AKT pathway. Additionally, cryptotanshinone can mediate the downregulation of PI3K/AKT pathway and induce apoptosis in breast cancer MCF-7 cells by activating GPER [31].

3.5 Extracellular signal-regulated kinase (ERK) signaling pathway

Hepatocellular carcinoma (HCC) is a malignant tumor, whose cell lines HCCLM3 and SMMC-7721 express high levels of GPER. The research by Qiu et al. [32] presented that compared with GPER-negative patients, GPER-positive HCC patients were closely related to the following factors, such as female, negative hepatitis B surface antigen, smaller tumor size, lower serum alpha-fetoprotein level and longer overall survival, and treatment of HCC cell lines HCCLM3 and SMMC-7721 with GPER-specific agonist G1 resulted in the activation of epidermal growth factor receptor (EGFR)/ERK and EGFR/AKT signaling pathways. In a murine Crohn's disease model induced by trinitrobenzene sulfonic acid, activation of GPER exerts the anti-inflammatory effects and lowers the mortality of mouse and C-reactive protein level, and this activation is accompanied by regulating ERK signaling pathway [33].

4. Physiological effects of GPER

4.1 GPER roles in inflammation

In a mouse model, the GPER selective agonist G-1 was found to reduce circulating levels of inflammatory cytokines tumor necrosis factor α (TNFα), monocyte chemoattractant protein-1 (MCP-1), and IL-6 and to lower the expression of inflammatory genes in a variety of metabolic tissues, when administered over a long period of time [34]. In microglia, GPER is involved in the anti-inflammatory process when genistein was administered to inhibit lipopolysaccharide (LPS)-induced microglia activation [35].

4.2 GPER roles in cardiovascular system

The blood pressure of male Sprague-Dawley (SD) rats can be significantly lowered by GPER1 activation with G1 [36]. Blood pressure is lowered in ovariectomy (OVX) SD rats and OVX mRen2 Lewis rats after chronic systemic administration with G-1 [37, 38]. The downstream of GPER1 activation involves signaling pathways such as nitric oxide (NO), which can mediate vasodilation occurrence and is a well-known vasodilator [39–41]. Downstream of GPER1 activation is also involved in renin angiotensin aldosterone system (RAAS), which controls blood pressure and glomerular filtration rate through renin and angiotensin I, II and aldosterone [42–44].

It is GPER1, while not ERα and ERβ, plays an important role in estrogen-responded heart rate regulation [45, 46]. GPER1 activation has a cardioprotective effect on heart failure in male mice and myocardial inflammation in male spontaneously hypertensive rats (SHRs), adriamycin-induced cardiotoxicity in male rats can be alleviated by GPER1 activation, and myocardial cell death can also be protected by GPER1 activation [47–50]. Moreover, in stress-induced cardiomyopathy, also known as Takotsubo syndrome, GPER activation exerts a protective role through balancing β2 adrenergic receptor (β2AR) with the Gαs and Gαi signaling pathways [51].

In short, it is worth noting that the current studies on the physiological effects of GPER are mainly from GPER knockout (KO) mice, which may have certain limitations. Therefore, when using GPER KO mice to study the mechanism of GPER-mediated action, the possibility of systemic compensation in mice should be taken into account [52].

4.3 GPER roles in stress

Endoplasmic reticulum stress (ERS) has been linked to several diseases such as cancer and diabetes and may be one of the possible inducers of pathological cell death and dysfunction [53, 54]. Icariin, an active constituent of epimedium, can inhibit ERS pathway through promoting the expression of GPER in diabetic kidney disease, and GPER is negatively related to the expression of endoplasmic reticulum response stress protein [55]. ERS pathways can be activated by GPER agonist, which leads to cancer cell death [56, 57]. GPER agonist G-1 can inhibit the increase of colonic crypt cell apoptosis in the colitis model [58] and improve epithelial cells after intestinal injury [59], and the protective effect of GPER is at least partially correlated with the inhibition of ERS.

Oxidative stress refers to the imbalance between oxidation and antioxidant effects in the body. Oxidative stress can occur when overexposed to reactive oxygen and nitrogen species, leading to damage of proteins, lipids, and DNA [60, 61]. Icariin also inhibits oxidative stress and promotes experimental diabetic nephropathy repair through GPER-mediated degradation of Keap1 protein and activation of Nrf2 protein [62]. Studies using a male rat model with traumatic brain injury-induced liver injury found that oxidative stress was significantly inhibited by 17β-estradiol [63–65]. In human skin fibroblasts, keratinocytes, and hepatic cells' 17 β-estradiol is also involved in inhibiting oxidative stress through the participation of GPER [66–68].

The range of cytoskeletal variation can be modulated by GPER, which is thought to be involved in mechanical transduction [69, 70]. When mechanical stress-mediated chondrocyte apoptosis occurs, chondrocyte apoptosis in osteoarthritis can be attenuated by GPER *via* causing the inhibition of Piezo 1 protein [71].

5. GPER in cancer

5.1 GPER in gynecological tumors

Bubb et al. [15] found that GPER is expressed in hepatocellular, pancreatic, renal and endometrial cancers, pancreatic neuroendocrine tumors, and pheochromocytomas by using GPER antibody 20H15L21. GPER is found to be expressed in 50–60% of breast cancer tissues [72–74], and is thought to be involved in the development of tamoxifen resistance in ERα-positive breast cancer [75, 76]. GPER can inhibit YAP1 phosphorylation by shutting down the Hippo pathway in breast cancer [27]. An immunohistochemical study of 1245 patients with primary invasive breast cancer showed that low expression of GPER was significantly related not only to clinicopathological and molecular characteristics of invasive behavior but also to poor survival in patients with breast cancer [77]. Chan et al. [78] showed that GPER induces phosphorylation of PKA and BAD-Ser118 to maintain breast cancer stem cells through activation of its ligands, which include tamoxifen (TMX) and GPER agonists increase cell proliferation and the number of breast cancer stem cells through the p-PKA/p-BAD signaling pathway.

GPER is expressed in cell lines such as human ovarian cancer cell line SKOV3, OVCAR-3, and OVCAR5 [79, 80], in both malignant and benign ovarian tumors, and overexpressed in some malignant ovarian tumors [81]. GPER is found in serous and mucinous ovarian adenocarcinoma biopsies. G-1 controls ovarian cancer cell proliferation, inhibits G2/M cell cycle progression, and promotes ovarian cancer cell apoptosis [79]. Zhu et al. showed that nuclear GPER is an independent negative marker for the prognosis of ovarian cancer patients, particularly for patients with highly malignant ovarian cancer [82].

Fujiwara et al. [83, 84] found that high GPER expression is related to poor prognosis in ovarian cancer patients, and lower GPER expression was positively correlated with overall survival time. However, in contrast to the results mentioned earlier, the study by Ignatov et al. found that high GPER expression positively affected 2-year disease-free survival in ovarian cancer patients [79]. The relation between GPER and ovarian cancer needs more studies to clarify.

5.2 GPER in tumor of male reproductive system

Concentrations of 17B-estradiol (E2) in the testis are 10 to 100 times higher than in the blood [85]. Although the molecular mechanisms by which estrogen and its receptors play a role in spermatogenesis are not fully understood, it is no doubt that estrogen is crucial in spermatogenesis [86]. GPER regulates the balance between testosterone and estrogen and is associated with the male reproductive system. GPER is predominantly located in the basal epithelium and mediates estrogen action, and the level of GPER expression is low in prostate tumor cells and in benign prostate tissue [87]. The expression of GPER was negatively correlated with the differentiation of prostate tumor cells [88]. In addition, it has been reported that GPER is a tumor suppressor that activates ERK1/2 and C-Jun/C-FOS signaling and thus inhibits PC3 cells, a cell line characteristic of prostatic small cell carcinoma and from arresting in G2 phase [89]. Lau et al. reported that GPER-specific agonist G1 binds to GPER to maintain ErK1/2 activation, thereby inhibiting prostate cancer cell growth and regulating metastasis characteristics [90].

Testicular germ cell carcinoma (TGCC) is a malignant solid tumor and is account for the main reason of death in young men. 17β-estradiol inhibits the proliferation of human seminoma cell lines by an ERβ-dependent mechanism, which is stimulated *in vitro* by GPER activation of ERK1/2 and protein kinase A. GPER is not expressed in non-seminoma, whereas GPER is exclusively overexpressed in seminoma, which is linked to the downregulation of ERβ, and may also be associated with genetic variations such as single-nucleotide polymorphisms [91].

It is worth mentioning that, regarding the potential sex-specific effects of GPER-1, there was no significant prognostic value of GPER-1 in the group of male breast cancers, which is contrary to what was expected in the studies of female breast cancers [92].

5.3 GPER in other cancer

Liu et al. [93–95] demonstrated that both cytoplasm-GPER (80.49%) and nucleus-GPER (53.05%) were detected in non-small-cell lung cancer (NSCLC) samples through immunohistochemical analysis, GPER activation promotes the growth of NSCLC cells *via* YAP1/QKI/circNOTCH1/m6A-methylated NOTCH1 pathway, and GPER antagonist G15 administration can reverse estrogen-induced progress of NSCLC cells, and these results suggest that blocking GPER signaling by G15 may be a new therapeutic target in NSCLC. In addition, the role of GPER in adrenal cortical carcinoma is related to mitochondria-related signal transduction, and GPER positively regulates mitochondrial apoptotic pathways through the EGR-1 pathway and ROS/EGR-1/BAX pathway, thereby inhibiting adrenocortical cancer cell growth [96].

6. GPER in retinal disease

Retinopathy of prematurity (ROP), characterized by abnormal growth of immature retinal blood vessels, is one of the leading causes of blindness and visual impairment in children worldwide. High incidence of ROP is intimately related to improved perinatal life care and poor ophthalmological special management in many developing countries, which lifts the survival rate of infants with ROP, and simultaneously, total infants with ROP [97].

Our research group has studied the role of GPER in ROP. The expression of GPER was found in endoplasmic reticulum of mouse retinal microglia, retinal ganglion cells, and retinal astrocytes. By GPER activation, hyperoxia-induced autophagy and apoptosis in these cells were reduced markedly, hyperoxia-induced decrease of the viability in these cells was reversed significantly, the high activity of inositol-1,4, 5-triphosphate receptor in these cells was reduced significantly also, and simultaneously, high calcium levels induced by hyperoxia in the endoplasmic reticulum was lowered again in these cells. These changes were reversed after the use of GPER antagonist G15. Thus, ERS irritated by hyperoxia was mitigated by G1 administration. These results suggest that GPER agonists may have therapeutic potential in the early stage of ROP [98–100].

7. Future directions

The key scientific issues that GPER needs to be further addressed as a key node are as follows: (1) the relationship between drugs (such as phytoestrogens, active ingredients of phytochemicals) and GPER and its signaling pathways needs to be systematically studied; (2) the interaction parameters and kinetics between GPER and ligands need to be further studied; (3) more work needs to be done to fully understand the potential of GPER1 in health regulatory mechanisms; and (4) the relationship between GPER and estrogen-related cancers, and the research on GPER as a therapeutic target for estrogen-related cancers needs to be carried out, which may unveil new therapies aimed at improving clinical outcomes of the diseases related to estrogen.

Acknowledgements

This work was supported by grants from the Key Project of Natural Science Basic Research Plan of Shaanxi Province (China, No. 2021JZ-60).

Conflict of interest

The authors declare no conflict of interest.

Abbreviations

EGFR	epidermal growth factor receptor
E2	17-β-estradiol
ER	estrogen receptor
GPCR	G protein-coupled receptor
GPER	G protein-coupled estrogen receptor
GPR30	G protein-coupled receptor 30
HCC	hepatocellular carcinoma
KO	knockout
SHR	spontaneously hypertensive rats
SD rat	Sprague-Dawley rats
SERM	selective estrogen receptor modulator
TGCC	testicular germ cell cancers
NSCLC	non-small-cell lung cancer

Author details

Hong-Bing Zhang*, Yao Wang and Bing Wang
Shaanxi Provincial Key Lab of Ophthalmology, Shaanxi Institute of Ophthalmology,
First Hospital of Xi'an, First Affiliated Hospital of Northwest University,
Xi'an, Shaanxi, China

*Address all correspondence to: zhanghongbing01@163.com

References

[1] Filardo EJ, Thomas P. Minireview: G protein-coupled estrogen receptor-1, GPER-1: Its mechanism of action and role in female reproductive cancer, renal and vascular physiology. Endocrinology. 2012;**153**(7):2953-2962. DOI: 10.1210/en.2012-1061

[2] Insel PA, Sriram K, Gorr MW, Wiley SZ, Michkov A, Salmerón C, et al. GPCRomics: An approach to discover GPCR drug targets. Trends in Pharmacological Sciences. 2019;**40**(6): 378-387. DOI: 10.1016/j.tips.2019.04.001

[3] Sriram K, Insel PA. G protein-coupled receptors as targets for approved drugs: How many targets and how many drugs? Molecular Pharmacology. 2018;**93**(4):251-258. DOI: 10.1124/mol.117.111062

[4] Eyster KM. The estrogen receptors: An overview from different perspectives. Methods in Molecular Biology. 2016;**1366**:1-10. DOI: 10.1007/978-1-4939-3127-9_1

[5] Takada Y, Kato C, Kondo S, Korenaga R, Ando J. Cloning of cDNAs encoding G protein-coupled receptor expressed in human endothelial cells exposed to fluid shear stress. Biochemical and Biophysical Research Communications. 1997;**240**(3):737-741. DOI: 10.1006/bbrc.1997.7734

[6] Carmeci C, Thompson DA, Ring HZ, Francke U, Weigel RJ. Identification of a gene (GPR30) with homology to the G-protein-coupled receptor superfamily associated with estrogen receptor expression in breast cancer. Genomics. 1997;**45**(3):607-617. DOI: 10.1006/geno.1997.4972

[7] Revankar CM, Cimino DF, Sklar LA, Arterburn JB, Prossnitz ER. A transmembrane intracellular estrogen receptor mediates rapid cell signaling. Science. 2005;**307**(5715):1625-1630. DOI: 10.1126/science.1106943

[8] Filardo EJ, Quinn JA, Frackelton AR Jr, Bland KI. Estrogen action via the G protein-coupled receptor, GPR30: Stimulation of adenylyl cyclase and cAMP-mediated attenuation of the epidermal growth factor receptor-to-MAPK signaling axis. Molecular Endocrinology. 2002;**16**(1):70-84. DOI: 10.1210/mend.16.1.0758

[9] Filardo E, Quinn J, Pang Y, Graeber C, Shaw S, Dong J, et al. Activation of the novel estrogen receptor G protein-coupled receptor 30 (GPR30) at the plasma membrane. Endocrinology. 2007;**148**(7):3236-3245. DOI: 10.1210/en.2006-1605

[10] Ariazi EA, Brailoiu E, Yerrum S, Shupp HA, Slifker MJ, Cunliffe HE, et al. The G protein-coupled receptor GPR30 inhibits proliferation of estrogen receptor-positive breast cancer cells. Cancer Research. 2010;**70**(3):1184-1194. DOI: 10.1158/0008-5472.Can-09-3068

[11] O'Dowd BF, Nguyen T, Marchese A, Cheng R, Lynch KR, Heng HH, et al. Discovery of three novel G-protein-coupled receptor genes. Genomics. 1998;**47**(2):310-313. DOI: 10.1006/geno.1998.5095

[12] Olde B, Leeb-Lundberg LM. GPR30/GPER1: Searching for a role in estrogen physiology. Trends in Endocrinology and Metabolism. 2009;**20**(8):409-416. DOI: 10.1016/j.tem.2009.04.006

[13] Isensee J, Meoli L, Zazzu V, Nabzdyk C, Witt H, Soewarto D, et al. Expression pattern of G protein-coupled

receptor 30 in LacZ reporter mice. Endocrinology. 2009;**150**(4):1722-1730. DOI: 10.1210/en.2008-1488

[14] Funakoshi T, Yanai A, Shinoda K, Kawano MM, Mizukami Y. G protein-coupled receptor 30 is an estrogen receptor in the plasma membrane. Biochemical and Biophysical Research Communications. 2006;**346**(3):904-910. DOI: 10.1016/j.bbrc.2006.05.191

[15] Bubb M, Beyer A-SL, Dasgupta P, Kaemmerer D, Sänger J, Evert K, et al. Assessment of G protein-coupled oestrogen receptor expression in normal and neoplastic human tissues using a novel rabbit monoclonal antibody. International Journal of Molecular Sciences. 2022 May 6;**23**(9):5191. DOI: 10.3390/ijms23095191

[16] Prossnitz ER, Arterburn JB. International union of basic and clinical pharmacology. XCVII. G protein-coupled estrogen receptor and its pharmacologic modulators. Pharmacological Reviews. 2015;**67**(3):505-540. DOI: 10.1124/pr.114.009712

[17] Barton M, Prossnitz ER. Emerging roles of GPER in diabetes and atherosclerosis. Trends in Endocrinology and Metabolism. 2015;**26**(4):185-192. DOI: 10.1016/j.tem.2015.02.003

[18] Dennis MK, Field AS, Burai R, Ramesh C, Petrie WK, Bologa CG, et al. Identification of a GPER/GPR30 antagonist with improved estrogen receptor counterselectivity. The Journal of Steroid Biochemistry and Molecular Biology. 2011;**127**(3-5):358-366. DOI: 10.1016/j.jsbmb.2011.07.002

[19] Dennis MK, Burai R, Ramesh C, Petrie WK, Alcon SN, Nayak TK, et al. In vivo effects of a GPR30 antagonist. Nature Chemical Biology. 2009;**5**(6): 421-427. DOI: 10.1038/nchembio.168

[20] Bologa CG, Revankar CM, Young SM, Edwards BS, Arterburn JB, Kiselyov AS, et al. Virtual and biomolecular screening converge on a selective agonist for GPR30. Nature Chemical Biology. 2006;**2**(4):207-212. DOI: 10.1038/nchembio775

[21] Albanito L, Madeo A, Lappano R, Vivacqua A, Rago V, Carpino A, et al. G protein-coupled receptor 30 (GPR30) mediates gene expression changes and growth response to 17beta-estradiol and selective GPR30 ligand G-1 in ovarian cancer cells. Cancer Research. 2007;**67**(4):1859-1866. DOI: 10.1158/0008-5472.Can-06-2909

[22] Di Mattia RA, Mariángelo JIE, Blanco PG, Jaquenod De Giusti C, Portiansky EL, Mundiña-Weilenmann C, et al. The activation of the G protein-coupled estrogen receptor (GPER) prevents and regresses cardiac hypertrophy. Life Sciences. 2020;**242**:117211. DOI: 10.1016/j.lfs.2019.117211

[23] Groban L, Tran QK, Ferrario CM, Sun X, Cheng CP, Kitzman DW, et al. Female heart health: Is GPER the missing link? Frontiers in Endocrinology (Lausanne). 2019;**10**:919. DOI: 10.3389/fendo.2019.00919

[24] Zhao Y, Yang Z, Miao Y, Fan M, Zhao X, Wei Q, et al. G protein-coupled estrogen receptor 1 inhibits the epithelial-mesenchymal transition of goat mammary epithelial cells via NF-κB signalling pathway. Reproduction in Domestic Animals. 2021;**56**(8):1137-1144. DOI: 10.1111/rda.13957

[25] Okamoto M, Suzuki T, Mizukami Y, Ikeda T. The membrane-type estrogen receptor G-protein-coupled estrogen receptor suppresses lipopolysaccharide-induced interleukin 6 via inhibition of nuclear factor-kappa B pathway

in murine macrophage cells. Animal Science Journal. 2017;**88**(11):1870-1879. DOI: 10.1111/asj.12868

[26] Zhou X, Wang S, Wang Z, Feng X, Liu P, Lv XB, et al. Estrogen regulates Hippo signaling via GPER in breast cancer. The Journal of Clinical Investigation. 2015;**125**(5):2123-2135. DOI: 10.1172/jci79573

[27] Deng Q, Jiang G, Wu Y, Li J, Liang W, Chen L, et al. GPER/Hippo-YAP signal is involved in Bisphenol S induced migration of triple negative breast cancer (TNBC) cells. Journal of Hazardous Materials. 2018;**355**:1-9. DOI: 10.1016/j.jhazmat.2018.05.013

[28] Diao W, Qian Q, Sheng G, He A, Yan J, Dahlgren RA, et al. Triclosan targets miR-144 abnormal expression to induce neurodevelopmental toxicity mediated by activating PKC/MAPK signaling pathway. Journal of Hazardous Materials. 2022;**431**:128560. DOI: 10.1016/j.jhazmat.2022.128560

[29] Yin H, Zhu Q, Liu M, Tu G, Li Q, Yuan J, et al. GPER promotes tamoxifen-resistance in ER+ breast cancer cells by reduced Bim proteins through MAPK/Erk-TRIM2 signaling axis. International Journal of Oncology. 2017;**51**(4):1191-1198. DOI: 10.3892/ijo.2017.4117

[30] Shi D, Zhao P, Cui L, Li H, Sun L, Niu J, et al. Inhibition of PI3K/AKT molecular pathway mediated by membrane estrogen receptor GPER accounts for cryptotanshinone induced antiproliferative effect on breast cancer SKBR-3 cells. BMC. Pharmacology and Toxicology. 2020;**21**(1):32. DOI: 10.1186/s40360-020-00410-9

[31] Shi D, Li H, Zhang Z, He Y, Chen M, Sun L, et al. Cryptotanshinone inhibits proliferation and induces apoptosis of breast cancer MCF-7 cells via GPER mediated PI3K/AKT signaling pathway. PLoS One. 2022;**17**(1):e0262389. DOI: 10.1371/journal.pone.0262389

[32] Qiu Y-A, Xiong J, Fu Q, Dong Y, Liu M, Peng M, et al. GPER-induced ERK signaling decreases cell viability of hepatocellular carcinoma. Frontiers in Oncology. 2021;**11**:638171. DOI: 10.3389/fonc.2021.638171

[33] Jacenik D, Zielińska M, Mokrowiecka A, Michlewska S, Małecka-Panas E, Kordek R, et al. G protein-coupled estrogen receptor mediates anti-inflammatory action in Crohn's disease. Scientific Reports. 2019;**9**(1):6749. DOI: 10.1038/s41598-019-43233-3

[34] Sharma G, Hu C, Staquicini DI, Brigman JL, Liu M, Mauvais-Jarvis F, et al. Preclinical efficacy of the GPER-selective agonist G-1 in mouse models of obesity and diabetes. Science Translational Medicine. 2020;**12**(528). DOI: 10.1126/scitranslmed.aau5956

[35] Du ZR, Feng XQ, Li N, Qu JX, Feng L, Chen L, et al. G protein-coupled estrogen receptor is involved in the anti-inflammatory effects of genistein in microglia. Phytomedicine. 2018;**43**:11-20. DOI: 10.1016/j.phymed.2018.03.039

[36] Haas E, Bhattacharya I, Brailoiu E, Damjanović M, Brailoiu GC, Gao X, et al. Regulatory role of G protein-coupled estrogen receptor for vascular function and obesity. Circulation Research. 2009;**104**(3):288-291. DOI: 10.1161/circresaha.108.190892

[37] Gohar EY, Daugherty EM, Aceves JO, Sedaka R, Obi IE, Allan JM, et al. Evidence for G-protein-coupled estrogen receptor as a pronatriuretic factor. Journal of the American Heart Association. 2020;**9**(10):e015110. DOI: 10.1161/jaha.119.015110

[38] Lindsey SH, Cohen JA, Brosnihan KB, Gallagher PE, Chappell MC. Chronic treatment with the G protein-coupled receptor 30 agonist G-1 decreases blood pressure in ovariectomized mRen2.Lewis rats. Endocrinology. 2009;**150**(8):3753-3758. DOI: 10.1210/en.2008-1664

[39] Lindsey SH, Liu L, Chappell MC. Vasodilation by GPER in mesenteric arteries involves both endothelial nitric oxide and smooth muscle cAMP signaling. Steroids. 2014;**81**:99-102. DOI: 10.1016/j.steroids.2013.10.017

[40] Tropea T, De Francesco EM, Rigiracciolo D, Maggiolini M, Wareing M, Osol G, et al. Pregnancy augments G protein estrogen receptor (GPER) induced vasodilation in rat uterine arteries via the nitric oxide—cGMP signaling pathway. PLoS One. 2015;**10**(11):e0141997. DOI: 10.1371/journal.pone.0141997

[41] Lindsey SH, da Silva AS, Silva MS, Chappell MC. Reduced vasorelaxation to estradiol and G-1 in aged female and adult male rats is associated with GPR30 downregulation. American Journal of Physiology. Endocrinology and Metabolism. 2013;**305**(1):E113-E118. DOI: 10.1152/ajpendo.00649.2012

[42] Gros R, Ding Q, Liu B, Chorazyczewski J, Feldman RD. Aldosterone mediates its rapid effects in vascular endothelial cells through GPER activation. American Journal of Physiology. Cell Physiology. 2013;**304**(6):C532-C540. DOI: 10.1152/ajpcell.00203.2012

[43] Ferreira NS, Cau SB, Silva MA, Manzato CP, Mestriner FL, Matsumoto T, et al. Diabetes impairs the vascular effects of aldosterone mediated by G protein-coupled estrogen receptor activation. Frontiers in Pharmacology. 2015;**6**:34. DOI: 10.3389/fphar.2015.00034

[44] Rigiracciolo DC, Scarpelli A, Lappano R, Pisano A, Santolla MF, Avino S, et al. GPER is involved in the stimulatory effects of aldosterone in breast cancer cells and breast tumor-derived endothelial cells. Oncotarget. 2016;7(1):94-111. DOI: 10.18632/oncotarget.6475

[45] Anderson JC, Beyger L, Guchardi J, Holdway DA. The effects of 17α-ethinylestradiol on the heart rate of embryonic Japanese medaka (*Oryzias latipes*). Environmental Toxicology and Chemistry. 2020;**39**(4):904-912. DOI: 10.1002/etc.4691

[46] Romano SN, Edwards HE, Souder JP, Ryan KJ, Cui X, Gorelick DA. G protein-coupled estrogen receptor regulates embryonic heart rate in zebrafish. PLoS Genetics. 2017;**13**(10):e1007069. DOI: 10.1371/journal.pgen.1007069

[47] Zhang X, Li T, Cheng HJ, Wang H, Ferrario CM, Groban L, et al. Chronic GPR30 agonist therapy causes restoration of normal cardiac functional performance in a male mouse model of progressive heart failure: Insights into cellular mechanisms. Life Sciences. 2021;**285**:119955. DOI: 10.1016/j.lfs.2021.119955

[48] Li WL, Xiang W, Ping Y. Activation of novel estrogen receptor GPER results in inhibition of cardiocyte apoptosis and cardioprotection. Molecular Medicine Reports. 2015;**12**(2):2425-2430. DOI: 10.3892/mmr.2015.3674

[49] Deschamps AM, Murphy E. Activation of a novel estrogen receptor, GPER, is cardioprotective in male and female rats. American Journal of Physiology. Heart and Circulatory

Physiology. 2009;**297**(5):H1806-H1813. DOI: 10.1152/ajpheart.00283.2009

[50] De Francesco EM, Rocca C, Scavello F, Amelio D, Pasqua T, Rigiracciolo DC, et al. Protective role of GPER agonist G-1 on cardiotoxicity induced by doxorubicin. Journal of Cellular Physiology. 2017;**232**(7):1640-1649. DOI: 10.1002/jcp.25585

[51] Fu L, Zhang H, Ong'achwa Machuki J, Zhang T, Han L, Sang L, et al. GPER mediates estrogen cardioprotection against epinephrine-induced stress. The Journal of Endocrinology. 2021;**249**(3):209-222. DOI: 10.1530/joe-20-0451

[52] Sharma G, Prossnitz ER. Targeting the G protein-coupled estrogen receptor (GPER) in obesity and diabetes. Endocrine and Metabolic Science. 2021;**2**. DOI: 10.1016/j.endmts.2021.100080

[53] Xu C, Bailly-Maitre B, Reed JC. Endoplasmic reticulum stress: Cell life and death decisions. The Journal of Clinical Investigation. 2005;**115**(10):2656-2664. DOI: 10.1172/JCI26373

[54] Shen X, Zhang K, Kaufman RJ. The unfolded protein response—A stress signaling pathway of the endoplasmic reticulum. Journal of Chemical Neuroanatomy. 2004;**28**(1-2):79-92. DOI: 10.1016/j.jchemneu.2004.02.006

[55] Su B, Cheng D, Chen G, Zhang S, Wang L, Wu X, et al. Icariin attenuation of diabetic kidney disease through inhibition of endoplasmic reticulum stress via G protein-coupled estrogen receptors. Journal of Biomedical Nanotechnology. 2022;**18**(2):488-497. DOI: 10.1166/jbn.2022.3242

[56] Vo DH, Hartig R, Weinert S, Haybaeck J, Nass N. G-protein-coupled estrogen receptor (GPER)-specific agonist G1 induces ER stress leading to cell death in MCF-7 cells. Biomolecules. 2019;**9**(9). DOI: 10.3390/biom9090503

[57] Lee SJ, Kim TW, Park GL, Hwang YS, Cho HJ, Kim JT, et al. G protein-coupled estrogen receptor-1 agonist induces chemotherapeutic effect via ER stress signaling in gastric cancer. BMB Reports. 2019;**52**(11):647-652. DOI: 10.5483/BMBRep.2019.52.11.007

[58] Wang Q, Li Z, Liu K, Liu J, Chai S, Chen G, et al. Activation of the G protein-coupled estrogen receptor prevented the development of acute colitis by protecting the crypt cell. The Journal of Pharmacology and Experimental Therapeutics. 2021;**376**(2):281-293. DOI: 10.1124/jpet.120.000216

[59] Chai S, Liu K, Feng W, Liu T, Wang Q, Zhou R, et al. Activation of G protein-coupled estrogen receptor protects intestine from ischemia/reperfusion injury in mice by protecting the crypt cell proliferation. Clinical Science (London, England). 2019;**133**(3):449-464. DOI: 10.1042/cs20180919

[60] Bowler RP, Crapo JD. Oxidative stress in allergic respiratory diseases. The Journal of Allergy and Clinical Immunology. 2002;**110**(3):349-356. DOI: 10.1067/mai.2002.126780

[61] Pizzino G, Irrera N, Cucinotta M, Pallio G, Mannino F, Arcoraci V, et al. Oxidative stress: Harms and benefits for human health. Oxidative Medicine and Cellular Longevity. 2017;**2017**:8416763, 10.1155/2017/8416763

[62] Wang K, Zheng X, Pan Z, Yao W, Gao X, Wang X, et al. Icariin prevents extracellular matrix accumulation and ameliorates experimental diabetic kidney disease by inhibiting oxidative stress

via GPER mediated p62-dependent Keap1 degradation and Nrf2 activation. Frontiers in Cell and Development Biology. 2020;**8**:559. DOI: 10.3389/fcell.2020.00559

[63] Amiresmaili S, Khaksari M, Shahrokhi N, Abolhassani M. Evolution of TLR4 role in mediating the hepatoprotective effects of estradiol after traumatic brain injury in male rats. Biochemical Pharmacology. 2020;**178**:114044. DOI: 10.1016/j.bcp.2020.114044

[64] Amiresmaili S, Shahrokhi N, Khaksari M, AsadiKaram G, Aflatoonian MR, Shirazpour S, et al. The Hepatoprotective mechanisms of 17β-estradiol after traumatic brain injury in male rats: Classical and non-classical estrogen receptors. Ecotoxicology and Environmental Safety. 2021;**213**:111987. DOI: 10.1016/j.ecoenv.2021.111987

[65] Amirkhosravi L, Khaksari M, Soltani Z, Esmaeili-Mahani S, Asadi Karam G, Hoseini M. E2-BSA and G1 exert neuroprotective effects and improve behavioral abnormalities following traumatic brain injury: The role of classic and non-classic estrogen receptors. Brain Research. 2021;**1750**:147168. DOI: 10.1016/j.brainres.2020.147168

[66] Savoia P, Raina G, Camillo L, Farruggio S, Mary D, Veronese F, et al. Anti-oxidative effects of 17 β-estradiol and genistein in human skin fibroblasts and keratinocytes. Journal of Dermatological Science. 2018;**92**(1):62-77. DOI: 10.1016/j.jdermsci.2018.07.007

[67] Surico D, Ercoli A, Farruggio S, Raina G, Filippini D, Mary D, et al. Modulation of oxidative stress by 17 β-estradiol and genistein in human hepatic cell lines in vitro. Cellular Physiology and Biochemistry. 2017;**42**(3):1051-1062. DOI: 10.1159/000478752

[68] Farruggio S, Raina G, Cocomazzi G, Librasi C, Mary D, Gentilli S, et al. Genistein improves viability, proliferation and mitochondrial function of cardiomyoblasts cultured in physiologic and peroxidative conditions. International Journal of Molecular Medicine. 2019;**44**(6):2298-2310. DOI: 10.3892/ijmm.2019.4365

[69] Lachowski D, Cortes E, Matellan C, Rice A, Lee DA, Thorpe SD, et al. G protein-coupled estrogen receptor regulates actin cytoskeleton dynamics to impair cell polarization. Frontiers in Cell and Development Biology. 2020;**8**:592628. DOI: 10.3389/fcell.2020.592628

[70] Wang Z, Sun L, Liang S, Liu ZC, Zhao ZY, Yang J, et al. GPER stabilizes F-actin cytoskeleton and activates TAZ via PLCβ-PKC and Rho/ROCK-LIMK-cofilin pathway. Biochemical and Biophysical Research Communications. 2019;**516**(3):976-982. DOI: 10.1016/j.bbrc.2019.06.132

[71] Sun Y, Leng P, Guo P, Gao H, Liu Y, Li C, et al. G protein coupled estrogen receptor attenuates mechanical stress-mediated apoptosis of chondrocyte in osteoarthritis via suppression of Piezo1. Molecular Medicine. 2021;**27**(1):96. DOI: 10.1186/s10020-021-00360-w

[72] Aiello F, Carullo G, Giordano F, Spina E, Nigro A, Garofalo A, et al. Identification of breast cancer inhibitors specific for G protein-coupled estrogen receptor (GPER)-expressing cells. ChemMedChem. 2017;**12**(16):1279-1285. DOI: 10.1002/cmdc.201700145

[73] Molina L, Figueroa CD, Bhoola KD, Ehrenfeld P. GPER-1/GPR30 a novel

estrogen receptor sited in the cell membrane: Therapeutic coupling to breast cancer. Expert Opinion on Therapeutic Targets. 2017;**21**(8):755-766. DOI: 10.1080/14728222.2017.1350264

[74] Jacenik D, Cygankiewicz AI, Krajewska WM. The G protein-coupled estrogen receptor as a modulator of neoplastic transformation. Molecular and Cellular Endocrinology. 2016;**429**:10-18. DOI: 10.1016/j.mce.2016.04.011

[75] Nass N, Kalinski T. Tamoxifen resistance: From cell culture experiments towards novel biomarkers. Pathology, Research and Practice. 2015;**211**(3):189-197. DOI: 10.1016/j.prp.2015.01.004

[76] Rondón-Lagos M, Villegas VE, Rangel N, Sánchez MC, Zaphiropoulos PG. Tamoxifen resistance: Emerging molecular targets. International Journal of Molecular Sciences. 2016;**17**(8). DOI: 10.3390/ijms17081357

[77] Martin SG, Lebot MN, Sukkarn B, Ball G, Green AR, Rakha EA, et al. Low expression of G protein-coupled oestrogen receptor 1 (GPER) is associated with adverse survival of breast cancer patients. Oncotarget. 2018;**9**(40):25946-25956. DOI: 10.18632/oncotarget.25408

[78] Chan Y-T, Lai ACY, Lin R-J, Wang Y-H, Wang Y-T, Chang W-W, et al. GPER-induced signaling is essential for the survival of breast cancer stem cells. International Journal of Cancer. 2020;**146**(6):1674-1685. DOI: 10.1002/ijc.32588

[79] Ignatov T, Modl S, Thulig M, Weißenborn C, Treeck O, Ortmann O, et al. GPER-1 acts as a tumor suppressor in ovarian cancer. Journal of Ovarian Research. 2013;**6**(1):51. DOI: 10.1186/1757-2215-6-51

[80] Yan Y, Jiang X, Zhao Y, Wen H, Liu G. Role of GPER on proliferation, migration and invasion in ligand-independent manner in human ovarian cancer cell line SKOV3. Cell Biochemistry and Function. 2015;**33**(8):552-559. DOI: 10.1002/cbf.3154

[81] Kolkova Z, Casslén V, Henic E, Ahmadi S, Ehinger A, Jirström K, et al. The G protein-coupled estrogen receptor 1 (GPER/GPR30) does not predict survival in patients with ovarian cancer. Journal of Ovarian Research. 2012;**5**:9. DOI: 10.1186/1757-2215-5-9

[82] Zhu CX, Xiong W, Wang ML, Yang J, Shi HJ, Chen HQ, et al. Nuclear G protein-coupled oestrogen receptor (GPR30) predicts poor survival in patients with ovarian cancer. The Journal of International Medical Research. 2018;**46**(2):723-731. DOI: 10.1177/0300060517717625

[83] Fujiwara S, Terai Y, Kawaguchi H, Takai M, Yoo S, Tanaka Y, et al. GPR30 regulates the EGFR-Akt cascade and predicts lower survival in patients with ovarian cancer. Journal of Ovarian Research. 2012;**5**(1):35. DOI: 10.1186/1757-2215-5-35

[84] Heublein S, Mayr D, Vrekoussis T, Friese K, Hofmann SS, Jeschke U, et al. The G-protein coupled estrogen receptor (GPER/GPR30) is a gonadotropin receptor dependent positive prognosticator in ovarian carcinoma patients. PLoS One. 2013;**8**(8):e71791. DOI: 10.1371/journal.pone.0071791

[85] Carreau S, Bourguiba S, Lambard S, Galeraud-Denis I, Genissel C, Levallet J. Reproductive system: Aromatase and estrogens. Molecular and Cellular Endocrinology. 2002;**193**(1-2):137-143. DOI: 10.1016/s0303-7207(02)00107-7

[86] Carreau S, Bouraima-Lelong H, Delalande C. Role of estrogens in

spermatogenesis. Frontiers in Bioscience (Elite Edition). 2012;**4**(1):1-11. DOI: 10.2741/356

[87] Rago V, Romeo F, Giordano F, Ferraro A, Carpino A. Identification of the G protein-coupled estrogen receptor (GPER) in human prostate: Expression site of the estrogen receptor in the benign and neoplastic gland. Andrology. 2016;**4**(1):121-127. DOI: 10.1111/andr.12131

[88] Xu S, Yu S, Dong D, Lee LTO. G protein-coupled estrogen receptor: A potential therapeutic target in cancer. Frontiers in endocrinology. 2019;**10**:725. DOI: 10.3389/fendo.2019.00725

[89] Chan QK, Lam HM, Ng CF, Lee AY, Chan ES, Ng HK, et al. Activation of GPR30 inhibits the growth of prostate cancer cells through sustained activation of Erk1/2, c-jun/c-fos-dependent upregulation of p21, and induction of G(2) cell-cycle arrest. Cell Death and Differentiation. 2010;**17**(9):1511-1523. DOI: 10.1038/cdd.2010.20

[90] Lau KM, Ma FM, Xia JT, Chan QKY, Ng CF, To KF. Activation of GPR30 stimulates GTP-binding of Gαi1 protein to sustain activation of Erk1/2 in inhibition of prostate cancer cell growth and modulates metastatic properties. Experimental Cell Research. 2017;**350**(1):199-209. DOI: 10.1016/j.yexcr.2016.11.022

[91] Chevalier N, Hinault C, Clavel S, Paul-Bellon R, Fenichel P. GPER and testicular germ cell cancer. Frontiers in Endocrinology. 2021;**11**:600404. DOI: 10.3389/fendo.2020.600404

[92] Maiwald JH, Sprung S, Czapiewski P, Lessel W, Scherping A, Schomburg D, et al. The impact of G protein-coupled oestrogen receptor 1 on male breast cancer: A retrospective analysis. Contemporary Oncology (Pozn). 2021;**25**(3):204-212. DOI: 10.5114/wo.2021.110010

[93] Liu C, Liao Y, Fan S, Tang H, Jiang Z, Zhou B, et al. G protein-coupled estrogen receptor (GPER) mediates NSCLC progression induced by 17β-estradiol (E2) and selective agonist G1. Medical Oncology. 2015;**32**(4):104. DOI: 10.1007/s12032-015-0558-2

[94] Liu C, Liao Y, Fan S, Fu X, Xiong J, Zhou S, et al. G-protein-coupled estrogen receptor antagonist G15 decreases estrogen-induced development of non-small cell lung cancer. Oncology Research. 2019;**27**(3):283-292. DOI: 10.3727/096504017x15035795904677

[95] Shen Y, Li C, Zhou L, Huang JA. G protein-coupled oestrogen receptor promotes cell growth of non-small cell lung cancer cells via YAP1/QKI/circNOTCH1/m6A methylated NOTCH1 signalling. Journal of Cellular and Molecular Medicine. 2021;**25**(1):284-296. DOI: 10.1111/jcmm.15997

[96] Casaburi I, Avena P, De Luca A, Sirianni R, Rago V, Chimento A, et al. GPER-independent inhibition of adrenocortical cancer growth by G-1 involves ROS/Egr-1/BAX pathway. Oncotarget. 2017;**8**(70):115609-115619. DOI: 10.18632/oncotarget.23314

[97] Romo-Aguas JC, González HLA, Meraz-Gutiérrez MP, Martínez-Castellanos MA. Retinopathy of prematurity: Incidence report of outliers based on international screening guidelines. International Journal of Retina Vitreous. 2019;**5**(Suppl. 1):53. DOI: 10.1186/s40942-019-0203-x

[98] Li R, Wang Y, Chen P, Meng J, Zhang H. G-protein coupled estrogen receptor activation protects the viability of hyperoxia-treated primary murine

retinal microglia by reducing ER stress. Aging (Albany NY). 2020;**12**(17):17367-17379. DOI: 10.18632/aging.103733

[99] Li R, Wang Y, Chen P, Meng J, Zhang H. Inhibiting endoplasmic reticulum stress by activation of G-protein-coupled estrogen receptor to protect retinal astrocytes under hyperoxia. Journal of Biochemical and Molecular Toxicology. 2021;**35**(2):e22641. DOI: 10.1002/jbt.22641

[100] Li R, Wang Y, Chen P, Meng J, Zhang H. G-protein-coupled estrogen receptor protects retinal ganglion cells via inhibiting endoplasmic reticulum stress under hyperoxia. Journal of Cellular Physiology. 2021;**236**(5): 3780-3788. DOI: 10.1002/jcp.30149

Section 2

Estrogen Action in the Oral and Olfactory Mucosae

Chapter 3

Effect of Hypoestrogenism on Oral Cavity

Pitu Wulandari

Abstract

Postmenopause is the period or period that occurs after women enter menopause. Menopause is the cessation of menstruation for 12 consecutive months. Menopause occurs due to the cessation of ovulation and hypoestrogenism or a decrease in the hormone estrogen, in this case, estradiol, a potent hormone owned by women. Hypoestrogenism causes various changes, including changes in the oral cavity. Some changes in the oral cavity are reduced salivary flow so that the mouth becomes dry. Dry mouth or xerostomia can increase the prevalence of caries in the oral cavity. In addition to decreased salivary flow, there is also a thinning of the gingival epithelium, which facilitates bacterial invasion and reduces alveolar bone mineral density associated with bone resorption, thereby increasing the risk of tooth mobility and even tooth loss. Prevention and treatment of periodontal destruction need to be done at this time, considering this is one of the factors that can improve the quality of life.

Keywords: postmenopause, hypoestrogenism, oral cavity, gingival, alveolar bone

1. Introduction

Postmenopause is the period after menopause in which the ovaries are no longer functioning. The level of the hormone estradiol is usually between 20 and 30 pg/ml [1, 2]. In postmenopausal, low estrogen hormone or hypoestrogenism causes androgens to maintain estrogen by changing androgens in peripheral fat (adipose) tissue so that the more subcutaneous fat a woman has, the higher the estrogen level [3]. After menopause, physiologically, women will enter the postmenopausal period; this period is marked by an increase in follicle-stimulating hormone (FSH) in the blood due to the ovaries being no longer active in producing hormones and the end of menstruation. This period begins with the end of menopause [4, 5].

The end of the menopausal transition is marked by amenorrhea (cessation of menstruation) at 60 days or more. The menstrual cycle at the end of the menopausal transition is characterized by increased cyclical variability, extreme hormone fluctuations, and an increased prevalence of anovulation. At this stage, FSH will increase, and sometimes the amount is the same as in the early reproductive years, which is associated with high estradiol levels. According to international standards and research data, the total FSH level is >25 IU/L at the end of the transition period or according to the standard pituitary gland; the average number is more than 40 IU/L [6, 7]. After menstruation stops for 12 months or enters menopause, women will undergo a postmenopausal period characterized by increased FSH in the blood and decreased estrogen hormone [5].

Factors that can affect menopause are as follows: (a) maternal age at menopause; (b) age at menarche; and (c) gestational age, and in this case, the women have first pregnancy at a later age has been related to the onset of menopause; (d) irregular menstrual cycles in women at forties as a marker of perimenopause or irregular cycles between the age 20 and 30 years had later age at menopause because prolongation of erratic menstrual functioning or delayed depletion of viable oocytes; (e) use of oral contraceptives; (f) number of pregnancies; (g) body mass index (BMI); (h) smoking and drinking alcohol; (i) physical activity; (j) serum lead levels; (k) polyunsaturated fat consumption; (l) socioeconomic status; (m) levels of education; (n) genetics; and (o) area of residence [8–10].

2. The process of postmenopause

Biological changes in women are broadly divided into two things: before entering the menstrual period and after the cessation of menstruation, which is directly influenced by hormonal changes. Hormones are specific molecules that regulate and potentially affect the development and integrity of tissues in the body. A woman's life is a complex biological journey from entering puberty to entering menopause. After menopause, women will experience a postmenopausal period [11, 12].

During puberty, sex hormone production will increase significantly during the reproductive period. After entering puberty, which is marked by the menstrual phase where there is an increase in the production of estrogen and progesterone secretions in a cyclic pattern that begins the onset of puberty, this cycle is called the reproductive or menstrual cycle [12]. Females change physically through the production of sex hormones at puberty. It begins with the secretion of increased dilation of blood vessels, thereby increasing permeability and increasing the production of prostaglandins by the anterior pituitary of gonadotropin hormones (follicle-stimulating hormone and luteinizing hormone), which causes the ovaries to initiate cyclical production and secretion of female sex hormones (estrogen and progesterone) [13].

Menopause is a biological condition due to ovarian failure, diagnosed when menstruation has stopped for 12 months. This condition cannot be explained through a pathological process, but it is associated with increased FSH in the blood of more than 40 IU/L; after that, women will experience a postmenopausal period. The physiology of menopause is the reduction of primary oocytes in the ovaries with atresia due to aging coupled with decreased sensitivity of the remaining oocytes to gonadotropins [14, 15]. Millions of women have undergone the menopausal transition; this period often has clinical implications that significantly affect women's health conditions. Changes during menopause occur due to hypoestrogenism, especially estradiol (E2), the most potent estrogen in the body. Estrogen and progesterone are responsible for the physiological changes of women in every phase of their life [16, 17].

Hormones fluctuate rapidly when women enter menopause. Changes in the mean values of FSH and estradiol indicate that FSH will continue to increase while estradiol will continue to decline for about 2 years. This is a critical factor in responding to symptoms related to the oral cavity. Raviraj et al. showed that oral symptoms were more common in postmenopausal women than in andropause men [18]. Female sex hormones and neuropathic factors are related to each other, so it is thought that neuropathic sensory fibers from the oral mucosa use discomfort in the oral cavity.

Women experienced more dental caries, abrasion, erosion, burning mouth sensation, and dentin hypersensitivity [2, 19].

3. Changing the oral cavity in postmenopausal

The mouth is a mirror of overall health, strengthening that oral health is an integral part of general health. In the elderly population, poor oral health has been considered a risk factor for public health problems. On the other hand, parents are more susceptible to disorders [20]. The oral cavity is a complex of many anatomical structures with different functions. It is made of soft and hard tissue, and the scope of its activity is vast. The oral cavity is susceptible to mechanical and chemical stimuli. Therefore, diseases in this area are often very unpleasant and burdensome to the patient. Awareness of oral health during the menopausal period is critical to minimize the inevitable discomfort due to hormonal changes [18, 21].

3.1 Burning mouth syndrome

Menopause initiates several physiological changes in women. One of them is a change in oral health. Discomfort in the oral cavity is found in menopausal women. Symptoms that often arise are dry mouth, pain in the mouth due to various causes, and burning mouth syndrome. Burning mouth syndrome is one of the main complications in menopause and postmenopause. It is a chronic condition characterized by a burning sensation of the oral mucosa, with or without dysgeusia (distortion of the sense of taste) and xerostomia in the absence of clinical lesions, laboratory abnormalities, or systemic causes [22]. The postmenopausal period affects the maturation of oral epithelial cells; this causes the epithelium to atrophy so that it is ultimately susceptible to inflammation complaints of the burning mouth that are often complained of in postmenopause. Some salivary functions are mediated by the level of expression of estrogen and estrogen receptors in the blood. Women also frequently complain of xerostomia associated with reduced estrogen [23].

Hormones fluctuate rapidly when women enter menopause. This is a key factor in responding to symptoms related to the oral cavity. Cell growth and function regulated by estrogen receptors, namely ERa and ER, are mediated by estrogen receptors. The soft tissues of the mouth and salivary glands will express ERβ receptors. This suggests that estrogen can play an important role in maintaining homeostasis in the oral cavity and salivary glands [18]. Inhibition of estrogen occurs after women complain of dry mouth due to decreased salivary secretion and sensation of burning of the mouth and tongue. Some women develop a condition known as menopausal gingivostomatitis, characterized by dry and shiny gingiva, bleeding easily, and varying in color from pale to erythematous [24, 25].

3.2 Xerostomia

Diseases of the oral mucosa often have a significant impact on the emotional state of the patients who experience them. Further aggravation of the unpleasant oral condition that accompanies menopause is associated with activation of the autonomic nervous system caused by chronic emotional anxiety. Oral health awareness during the menopausal period is critical to minimize the inevitable discomfort due to hormonal changes. Dry mouth or xerostomia is one of the most common symptoms women report in menopause. Xerostomia is a dry mouth condition most often caused by a decrease in saliva or when salivary factors are standard but with fewer components [21]. Besides hormonal factors, drugs such as antidepressants, antihistamines, antihypertensives, and diuretics, which are taken continuously in postmenopausal women, also cause reduced salivary flow [25]. Reduced salivary flow (intrinsic cleaning mechanism) and poor oral hygiene (extrinsic cleaning mechanism) can accumulate heavy plaque on the teeth and denture surfaces. It is a condition that predisposes to and increases periodontal disease and dental caries [20, 26].

3.3 Dental caries

Saliva plays an important role in defense of tissues in the oral cavity and avoiding foreign objects entering the body. When saliva flow is reduced, the tendency to experience disturbances in the oral cavity will increase [27]. Decreased secretion of saliva, a compelling defense of the oral cavity, can cause many problems, such as increased dental caries, oral infections, dysphagia, impaired taste, and increased mucosal sensitivity against mechanical injury. The normal flow of unstimulated and stimulated saliva is essential to ensure adequate and continuous lubrication of the oral tissues. The salivary fluid's characteristics are critical for forming a food bolus, dissolving taste substances, and transporting them to taste receptors. It also facilitates chewing, swallowing food, and talking. Insufficient saliva can disrupt the balance of microbes that benefit pathogens, such as *Candida albicans* and *Streptococcus mutans*, because a moist environment is also essential for the colonization and growth of microorganisms on the oral surface [21, 28, 29].

Salivary glands also depend on reproductive hormones, contributing to salivary secretion and consistency changes. This condition affects the teeth and increases the risk of caries. Postmenopausal women with hormone therapy have an improved quality of life as oral discomfort decreases. This is partly due to the increased secretion of salivary flow. Postmenopausal and premenopausal saliva composition appears to be estrogen-dependent. However, hormone therapy does not affect the total number of bacteria in saliva in premenopausal or postmenopausal [21, 30].

Rukmini et al.'s research showed a significant difference in dental caries between postmenopausal women and women who were not yet menopausal. The average number of caries sufferers in postmenopausal women was higher than those who had not experienced menopause [27]. Slowed secretory activity in postmenopause will facilitate bacterial and fungal colonization, which increases dental caries and increases the risk of pathogen access in the subepithelial mucosa due to the weak ability of saliva to perform self-cleaning The level of sex hormones in the saliva is also related to the number of hormones in the blood [31].

Hypoestogenism in postmenopausal affects the salivary flow rate, which at the same time also affects the condition of microorganisms in the oral cavity. The decreased salivary flow will affect the composition of salivary fluid and gingival crevicular fluid. Saliva is a factor that plays a role in maintaining the pH of the oral cavity by neutralizing acids from food and drink as well as bacterial activity to reduce the risk of periodontal disease [32].

3.4 Periodontal change

After menopause, women become more susceptible to periodontal disease. This is partly due to estrogen deficiency which causes gingival inflammation and bone resorption. Estrogen and androgen hormones are present in the periodontium, so postmenopausal hormone imbalances will affect the periodontium. The decrease in the hormone estrogen causes changes in immune function and the ecology of microorganisms in the body, especially in the oral cavity [33–35]. Microcirculation in the postmenopausal oral cavity shows that the diameter of the loops, tortuosity of blood vessels in the labial mucosa, and changes in the density of the periodontal mucosa predispose to periodontal inflammation [36].

Chronic disease conditions increase the sulcus depth, so food debris and periodontal pathogens accumulate in the sulcus. This is harmful to the periodontal ligament fibers attached to the gingiva and teeth. If the microbial accumulation persists in the sulcus for a long time, the bacteria will penetrate, and gingival inflammation occurs and ultimately destroys the connective tissue and periodontal ligament fibers. This process will continue, causing the sulcus to deepen and form a periodontal pocket [37, 38].

In postmenopause, changes in the oral cavity that occur due to hormonal changes are a picture of physiological aging in oral tissues, so this has the potential to cause periodontitis [33, 39]. The increase in postmenopausal *pro-inflammatory* cytokines causes changes in *anti-inflammatory cytokines*, which are associated with weakened monocyte and macrophage function due to estrogen deficiency. This is the beginning of the influence of menopause on tissue function [40, 41].

3.4.1 Gingival

The role of estrogen in periodontal tissue is that estrogen can stimulate the proliferation of gingival fibroblasts, stimulate the synthesis and maturation of the gingival connective tissue, increase gingival inflammation without increasing plaque, increase cellular proliferation in blood vessels, and reduce keratinization through an increase in epithelial glycogen, causing narrowing/shrinking of the epithelial barrier. It also decreases T cells that mediate inflammation, suppresses the number of leukocytes (leukopoiesis) in the bone marrow, stimulates phagocytosis of polymorphonuclear (PMN)), inhibits PMN chemotaxis, and inhibits pro-inflammatory cytokines released by the bone marrow [11, 42–44]. There are two theories for the action of steroid hormones on gingival cells: (a) changes in the effectiveness of the epithelial barrier against bacterial attack and (b) effects on collagen maintenance and repair. Estrogen receptors are located in the oral mucosa, so changes in this hormone will directly affect the oral cavity. Postmenopausal oral health affects general health and requires attention and other factors [18, 22, 45].

Steroid sex hormones have been shown to, directly and indirectly, influence cell proliferation, differentiation, and growth of target tissues, including keratinocytes and fibroblasts in the gingiva [46]. Different with the stimulatory effect of estrogen on gingival fibroblast proliferation, collagen and non-collagen protein production was reduced when physiological concentrations of estradiol were introduced to fibroblasts. Collagen reduction and non-collagen protein production occur by strained fibroblasts. Therefore, estrogen does not affect collagen synthesized by gingival fibroblasts [47].

3.4.2 Cementum and periodontal ligaments

The periodontal tissue's complex histological structure becomes affected during women's postmenopausal fluctuations in sex hormones. The components involved include the gingival epithelium and the underlying connective tissue as well as the extracellular matrix, the periodontal ligament, the alveolar bone and cementum of the tooth roots, the vascular endothelium, and circulating immune cells that explicitly reflect the various activations whose actions are systemically influenced by female sex hormones [48].

Menopause is a physiological phenomenon that occurs in aging women. Periodontal disease is related to changes that occur when entering menopause, one of which is alkaline phosphatase (ALP) which plays a role in the replacement of bone and periodontal tissue in general. Calcium and alkaline phosphatase (ALP) are bone markers associated with accelerated postmenopausal bone loss. Alkaline phosphatase forms the periodontal tissue's periodontal ligament, cementum, and homeostasis. During the acellular cementum apposition mechanism, vigorous ALP activity was found in the periodontal ligament, so postmenopausal ALP changes also affect periodontal conditions, including cementum [49, 50]. In menopause, the periodontium is directly affected by a neural mechanism against estrogen deficiency. The integrity of the periodontal tissue is regulated by several factors, including the ability of the periodontal ligament to differentiate into osteoblasts or cementoblasts. Hypoestrogenism substantially impacts the etiology, manifestation, and severity through the expression of osteoprotegerin and *receptor activator of nuclear factor kappa-B ligand* (RANKL) in periodontal ligament cells via ER [36, 51].

The significant role of estrogen in the differentiation of periodontal ligament stem cells (PDLSCs) is to exhibit hypoestrogenism on alveolar bone resorption during menopause. In human periodontal ligament cells, there is a decrease in collagen synthesis in fibroblasts due to the influence of estrogen levels. A gradual reduction in estrogen will induce a dependent intensification of procollagen production. Estrogen modulates the activity of target cells to bind to intracellular estrogen receptors. These hormone receptors will assist in the regulation of cell growth and cell differentiation in response to estrogen. When estrogen levels decrease, it will interfere with osteogenic differentiation of periodontal ligament cells, inhibit osteoblast differentiation of human periodontal ligament cells, and interfere with the formation of mineral nodules in the periodontal ligament [52, 53].

3.4.3 Alveolar bone

Periodontitis is a multifactorial infection characterized by a destructive inflammatory process in the supporting tissues of the teeth, causing the formation of periodontal pockets and alveolar bone resorption, which can lead to tooth loss. Periodontal bacteria are the main cause of periodontitis which initiates periodontitis; the host response also determines the progression and severity of the disease. Alteration of the immune response of the host and surrounding cells by periodontal pathogens and their virulence factors results in a complex network of pro- and anti-inflammatory cytokines that play a role in periodontal disease [54, 55].

In patients suffering from progressive periodontitis, the patient must accept the treatment plan given to them to prevent further periodontal destruction. Methods to detect periodontal disease can be done through a diagnostic process, namely interviews, patient history taking, clinical examination, and radiographic examination. The periodontal examination should be standard practice, since it is needed to identify patients at risk and rule out bacteria early on [37]. Low estrogen production after menopause is associated with increased production of interleukin 1 *(IL-1), IL-6, IL-8, IL-10,* tumor necrosis factor-alpha, granulocyte colony-stimulating factor, and macrophages-stimulating factor that stimulates mature osteoclasts, modulates bone cell proliferation, and induces skeletal and alveolar bone resorption. Age, smoking, socioeconomic factors, and systemic conditions can play essential roles as confounders of the relationship between menopause and tooth loss. The causes of missing teeth are complex but reflect the cumulative state of oral health over time [53, 56].

Swelling of endothelial cells and pericytes of the venules, attachment of granulocytes and platelets to the vessel wall, and proliferation of blood vessels are associated with variations in estrogen levels. Decreased estrogen activity is the leading cause of decreased alveolar bone remodeling. Disruption of positive feedback on the hypothalamic–pituitary axis to the ovary leads to increased alveolar bone resorption [36, 57]. Osteoporosis is the most common cause of estrogen deficiency in postmenopausal women. The risk of osteoporosis and fracture is a dynamic process involving metabolism and bone remodeling that compromises the health of the periodontium [51].

Postmenopausal estrogen deficiency is a critical factor in the pathogenesis of osteoporosis and is involved in rapid bone resorption rather than bone formation and is associated with increased alveolar bone loss resulting in a high risk of tooth loss [53, 58]. Systemic factors are also responsible for postmenopausal osteoporotic bone loss combined with other factors such as periodontal disease, thereby increasing alveolar bone loss. The decrease in bone mineral density in menopause is related to the trabecular pattern and the speed of bone resorption, which predisposes to periodontal disease. The increased cytokine production and bone turnover make the host more susceptible to periodontal disease, so treatment is carried out to reduce the rate of alveolar bone loss and the progression of postmenopausal periodontal disease [58].

4. Conclusion

Diseases of the oral cavity, especially periodontitis and dental caries, are best diagnosed early, so treatment can be started more quickly and tooth loss can be prevented. Periodontal therapy in postmenopausal will be successful if the clinician understands the disease process to establish the right diagnosis to maintain the patient's oral health. Estrogen is the potential hormone that participates in maintaining stability of their oral health, especially in postmenopausal people who are prone to experiencing diseases in the oral cavitiy due to changes in reproductive hormones. Adequate oral care is one thing that can improve postmenopausal quality of life.

Author details

Pitu Wulandari
Department of Periodontics, Universitas Sumatera Utara, Medan, Indonesia

*Address all correspondence to: pitu.wulandari@usu.ac.id

References

[1] Baziad A. Menopause, Andropause dan Terapi Sulih Hormon (TSH). In: Menopause dan Andropause. Jakarta: Yayasan Bina Pustaka Sarwono Prawirorahardjo; 2003. pp. 1-6

[2] Sakhalkar M. Awareness of middle aged women about changes and it's effect on health during menopause. Research Journal. 2015;**4**(3):1-5

[3] Taketani Y, Yano T, Kugu K. The climacteric as a crucial stage of female life. Journal of Japan Medical Association. 2004;**128**(9):166-170

[4] Charles C et al. Effects of high-dose isoflavones on metabolic and inflammatory markers in healthy postmenopausal women. Menopause. 2009;**16**(2):395-400. DOI: 10.1097/gme.0b013e3181857979

[5] Rodríguez-landa JF, Puga-olguín A, Germán-ponciano LJ. Anxiety in Natural and Surgical Menopause – Physiologic and Therapeutic Bases. Intech. 2015: 174-190. DOI: 10.5772/60621

[6] Harlow SD et al. Executive summary of the stages of reproductive aging workshop + 10: Addressing the unfinished agenda of staging reproductive aging. The Journal of Clinical Endocrinology and Metabolism. 2012;**97**(4):1159-1168. DOI: 10.1210/jc.2011-3362

[7] Stricker R et al. Establishment of detailed reference values for luteinizing hormone, follicle stimulating hormone, estradiol, and progesterone during different phases of the menstrual cycle on the abbott architect analyzer. Clinical Chemistry and Laboratory Medicine. 2006;**44**(7):883-887. DOI: 10.1515/CCLM.2006.160

[8] Ceylan B, Özerdoğan N. Factors affecting age of onset of menopause and determination of quality of life in menopause. Journal of Turkish Society Obstetrics and Gynecology. 2015;**12**(1):43-49. DOI: 10.4274/tjod.79836

[9] Parazzini F. Determinants of age at menopause in women attending menopause clinics in Italy. Menopause Journal. 2006;**56**:280-287. DOI: 10.1016/j.maturitas.2006.09.003

[10] Bromberger JT, Matthews KA, Kuller LH, Wing RR, Meilahn EN, Plantinga P. Prospective study of the determinants of age at menopause. American Journal of Epidemiology. 1997;**145**:124-133

[11] Güncü G, Tözüm T, Çaglayan F. Effects of endogenous sex hormones on the periodontium—Review of literature. Australian Dental Journal. 2005;**50**(3):138-145. DOI: 10.1111/j.1834-7819.2005.tb00352.x

[12] Rothmund WL, O'Kelley-Wetmore AD, Jones ML, Smith MB. Oral manifestations of menopause: An interprofessional intervention for dental hygiene and physician assistant students. Journal of Dental Hygiene. 2017;**91**(6):21-32

[13] Amar S, Chung KM. Influence of hormonal variation on the periodontium in women. Periodontology 2000. 2000;**6**(1):79-87. DOI: 10.1111/j.1600-0757.1994.tb00028.x

[14] Archer DF et al. Menopausal hot flushes and night sweats: Where are we now ? Climacteric. 2011;**4**:515-528. DOI: 10.3109/13697137.2011.608596

[15] Soules MR et al. Executive summary: Stages of reproductive aging workshop (STRAW). Fertility and Sterility. 2001;**76**(5):874

[16] Yıldırım TT, Kaya FA. The effects of menopause on periodontal tissue. International Dental Reseaarch. 2011;**3**(2):81-66

[17] Gold EB. The timing of the age at which natural menopause occurs. Obstetrics and Gynecology Clinics of North America. 2011;**38**(3):425-440. DOI: 10.1016/j.ogc.2011.05.002

[18] Raviraj J et al. Influence of menopause on oral health: A cross sectional study. International Journal of Health Sciences and Research. 2016;**6**(2):232-235. Available from: www.ijhsr.org

[19] Mutneja P, Dhawan P, Raina A, Sharma G. Menopause and the oral cavity. Indian Journal of Endocrinology Metabolism. 2012;**16**(4):548. DOI: 10.4103/2230-8210.98007

[20] Bhadbhade S. Aging & periodontium. International Journal of Dental and Oral Science. 2015;**6**:79-83

[21] Ciesielska A, Kusiak A, Ossowska A, Grzybowska ME. Changes in the oral cavity in menopausal women—A narrative review. International Journal of Environmental Research and Public Health. 2022;**19**:1-12. DOI: 10.3390/ijerph19010253

[22] Dutt P, Chaudhary S, Kumar P. Oral health and menopause: A comprehensive review on current knowledge and associated dental management. Annals of Medical and Health Sciences Research. 2013;**3**(3):320-323. DOI: 10.4103/2141-9248.117926

[23] Välimaa H et al. Estrogen receptor is the predominant estrogen receptor subtype in human oral epithelium and salivary glands. The Journal of Endocrinology. 2004;**180**:55-62

[24] Singh M et al. Periodontal status in pre-and post-menopausal women: A review. Asian Pacific Journal of Health Science. 2018;**5**(2):136-141. DOI: 10.21276/apjhs.2018.5.2.26

[25] Loesche WJ, Lopatin DE. Interactions between periodontal disease, medical diseases and immunity in the older individual. Periodontology 2000. 2000;**16**:80-105. DOI: 10.1111/j.1600-0757.1998.tb00117.x

[26] Sreebny LM, Schwartz SS. A reference guide to drugs and dry mouth. Gerodontology. 1986;**5**(2):75-79. DOI: 10.1111/j.1741-2358.1997.00033.x

[27] Rukmini JN, Sachan R, Sibi N, Meghana A, Malar CI. Effect of menopause on saliva and dental health. Journal of International Society. 2018;**8**:529-533. DOI: 10.4103/jispcd.JISPCD

[28] Meurman JH, Tarkkila L, Tiitinen A. The menopause and oral health. Maturitas. 2009;**63**(1):56-62. DOI: 10.1016/j.maturitas.2009.02.009

[29] Foglio-Bonda PL, Rocchetti V, Nardella A, Fantinato M, Sandhu K, Foglio-Bonda A. Salivary pH and flow rate in menopausal women. European Review for Medical and Pharmacological Sciences. 2019;**23**:918-922. DOI: 10.26355/eurrev_201902_16982

[30] Mojabi KB, Esfahani M, Hashemi HJ. Evaluation of unstimulated salivary flow rate and oral symptoms in menopausal women. Journal of Dental. 2007;**4**:918-103

[31] Bianchi S, Nottola SA, Torge D, Palmerini MG, Necozione S, Macchiarelli G. Association between female reproductive health and mancozeb: Systematic review of experimental models. International Journal of Environmental Research and Public Health. 2020;**17**:1-15. DOI: 10.3390/ijerph17072580

[32] Hernández-Vigueras S et al. Oral microbiota, periodontal status, and osteoporosis in postmenopausal females. Journal of Periodontology. 2016;**87**(2):124-133. DOI: 10.1902/jop.2015.150365

[33] Buencamino MCA, Palomo L, Thacker HL. How menopause affects oral health, and what we can do about it. Clinical Journal of Medicine. 2009;**8**:124-133

[34] Liu JM et al. Clinical investigations relationships between the changes of serum levels of OPG and RANKL with age, menopause, bone biochemical markers and bone mineral density in chinese women aged 20-75. Calcified Tissue Internationa. 2004;**76**:1-6

[35] Qasim AA, Mt K, Nuaimy A, Ahmed A. Assessment of periodontal status among premenopausal and postmenopausal women in Mosul City. Journal of Education Science. 2012;**25**(1):73-76

[36] Di Naro E, Loverro M, Converti I, Loverro MT, Ferrara E, Rapone B. The effect of menopause hypoestrogenism on osteogenic differentiation of periodontal ligament cells (PDLC) and stem cells (PDLCs): A systematic review. Health. 2021;**925**:1-11

[37] Solanki G. A general overview of periodontal pockets. International Journal of Biomedical Research. 2012;**3**:127-130

[38] Mariano FS, Sardi J d CO, Duque C, Höfling JF, Gonçalves RB. The role of immune system in the development of periodontal disease: A brief review. Rev Odonto Cienc. 2010;**25**(3):300-305

[39] Puspitadewi SR et al. The relation of follicle stimulating hormone and estrogen to mandibular alveolar bone resorption in postmenopausal women. Journal of International Dental and Medical Research. 2017;**10**(3):1-7

[40] Faizuddin M, Bharathi SH, Rohini NV. Estimation of interleukin-1β levels in the gingival crevicular fluid in health and in inflammatory periodontal disease. Journal of Periodontal Research. 2003;**38**:111-114

[41] Malutan AM, Dan M, Nicolae C, Carmen M. Proinflammatory and anti-inflammatory cytokine changes related to menopause. Prz. Menopauzalny. 2014;**13**(3):162-168. DOI: 10.5114/pm.2014.43818

[42] Anand V, Gulati M, Anand B, Singh A, Daing A. Influence of hormones in periodontium—A review. Inernational Journal of Research. 2012;**1**:1-14

[43] Grover CM, More VP, Singh N, Grover S. Crosstalk between hormones and oral health in the mid-life of women: A comprehensive review. Journal of International Society. 2014;**4**(Supplement. 1):S5-S10. DOI: 10.4103/2231-0762.144559

[44] Georgescu DH. Histopathology of marginal superficial periodontium at menopause. International Journal of Medical Dental. 2012;**2**(1):63-68

[45] Suri V, Suri V. Menopause and oral health. Journal of Midlife Health. 2014;**5**(3):115-120. DOI: 10.4103/0976-7800.141187

[46] Meisel P, Reifenberger J, Haase R, Nauck M, Bandt C, Kocher T. Women

are periodontally healthier than men, but why don' t they have more teeth than men? Journal of North American Menopause Society. 2008;**15**(2):270-275. DOI: 10.1097/gme.0b013e31811ece0a

[47] Jafri Z, Bhardwaj A, Sawai M, Sultan N. Influence of female sex hormones on periodontium: A case series. Journal of National Science and Biology Medicine. 2015;**6**(1):S146-S149. DOI: 10.4103/0976-9668.166124

[48] Iliescu AA, Perlea P, Earar K. Insight in periodontal status of menopausal women. Romanian Journal of Oral Rehabilitation. 2020;**12**(2):250-257

[49] Najeeb S, Zafar MS, Khurshid Z, Zohaib S, Almas K. The role of nutrition in periodontal health: An update. Nutrients. 2016;**8**:1-18. DOI: 10.3390/nu8090530

[50] Hutomo DI, Masulili SLC, Tadjoedin FM, Kusdhany LS. Serum alkaline phosphatase and calcium in relation to periodontal status among perimenopausal and postmenopausal women. The Open Dentistry Journal. 2021;**15**(1):703-708. DOI: 10.2174/1874210602115010703

[51] Nastri L et al. Do dietary supplements and nutraceuticals have effects on dental implant osseointegration? A scoping review. Nutrients. 2020;**12**:1-16. DOI: 10.3390/nu12010268

[52] Bianchi S et al. Ultrastructural markers of quality are impaired in human metaphase II aged oocytes: A comparison between reproductive and in vitro aging. Journal of Assisted Reproduction and Genetics. 2015;**32**:1343-1358. DOI: 10.1007/s10815-015-0552-9

[53] Pacifici R. Estrogen, cytokines, and pathogenesis of postmenopausal osteoporosis. Journal of Bone Mineral Research. 1996;**11**(8):1043-1051. DOI: 10.1002/jbmr.5650110802

[54] Al-Ghurabei BH, Zahraa F, Al-hassan AA. Clinical relevance of IL-1B/IL -10 AND TNF-A/IL –10 ratio in chronic periodontitis patients. Internatiinal Journal of Recent Science and Research. 2013;**4**(6):275-277

[55] Puspitadewi SR, Kusdhany LS, Masulili SLC, Wulandari P, Iskandar HB, Auerkari EI. The role of parathyroid hormone in alveolar bone resorption on postmenopausal women. The Open Dentistry Journal. 2020;**14**:82-87. DOI: 10.2174/1874210602014010082

[56] Yakar N et al. Oral health and emotional well-being in premenopausal and postmenopausal women: A cross-sectional cohort study. BMC Women's Health. 2021;**21**:1-9. DOI: 10.1186/s12905-021-01480-5

[57] Cecoro G et al. The role of bone anabolic drugs in the management of periodontitis: A scoping review. European Cells & Materials. 2021;**41**:316-331. DOI: 10.22203/eCM.v041a20

[58] Deshpande N, Sharma D. Periodontal management of post menopausal women. Indian Journal of Multidisciplinary Dental. 2012;**2**(4):565-569

Chapter 4

Functional Roles of Estradiol in the Olfactory and Vomeronasal Mucosae of Mammals: A Working Hypothesis

Shigeru Takami and Sawa Horie

Abstract

It has been known that androgens and estrogens, which are referred to as sex steroids, make many effects on two major nasal chemosensory mucosae such as olfactory mucosa and vomeronasal organ. Our studies conducted in rodents have demonstrated that two of the constituent cells in the olfactory mucosa, sustentacular cells and acinar cells in the associated glands of the olfactory mucosa, Bowman's glands, express four different enzymes involved in the biosynthesis of estradiol-17β (E2). Furthermore, our ongoing study has shown that olfactory sensory cells contain immunoreactivity for an estrogen receptor (beta-type). In case of vomeronasal organ, vomeronasal sensory cells express two enzymes that catalyze conversion of E2 and estrone, and that of testosterone and androstenedione. In addition, vomeronasal sensory cells contain an estrogen receptor (alpha-type). These results strongly suggest that de novo synthesis of E2 and metabolism of E2 take place in the olfactory mucosa and vomeronasal organ, respectively. With special emphasis of subcellular characteristics of steroid-producing cells, such as presence of large amount of smooth endoplasmic reticulum and vesicular mitochondria, we will introduce our findings and present working hypotheses for E2 functions in the olfactory mucosa and vomeronasal organ.

Keywords: olfactory sensory cell, olfactory sustentacular cell, Bowman's glands, smooth vomeronasal sensory cell, smooth endoplasmic reticulum, estradiol-17β, estrogen receptor

1. Introduction

The major chemosensory mucosa in mammalian species is olfactory mucosa (OM). Olfactory sensory cells (OSCs) that detect odorants and are housed in the OM share not only common physiological and morphological features of sensory cells but also those of bipolar neurons [1–4]. Electrical signals induced by odorants are conducted through axons to the surface of main olfactory bulb (MOB) and transmitted into olfactory bulbar neurons, such as mitral and tufted cells via chemical synapses

[3]. In addition to the OM, many land-living mammals contain vomeronasal organs (VNOs) that are bilaterally located at the base of the nasal septum [5–8]. Sensory cells of the VNO, vomeronasal sensory cells (VSCs), which are bipolar neurons as well, mainly detect species-specific odorants including pheromones and transmit their odorant information to projection neurons in the accessory olfactory bulb, so-called mitral/tufted cells [9]. Other than the OM and VNO, septal organ and Gruenberg ganglion [8] contain chemosensory cells that have been found at least rodent species [10]. Locations of these four chemosensory mucosae and their primary brain centers are schematically shown in **Figure 1A**. In this chapter, we will focus on two of them, OM and VNO, to demonstrate experimental data using rats to suggest de novo synthesis of estradiol-17β (E2) in the OM and metabolizing of E2 in the VNO.

By a radioautographic technique, Stumpf and Sar [11] demonstrated the incorporation of 3H-labeled estradiol into cell nuclei of OSCs, duct cells of Bowman's glands, and mitral cells in the main olfactory bulb where synaptic terminals of OSCs are present. They suggested that these cells contain estrogen receptors and are influenced by estradiol. In ferrets, sex steroids including estradiol increased ferrets' responsiveness to low concentrations of odors emitted from anal scent glands [12]. Conspecific odor preference by male rats was influenced by estradiol [13]. Bakker et al. [14] demonstrated in aromatase knockout male mice that exogenous estrogens restore

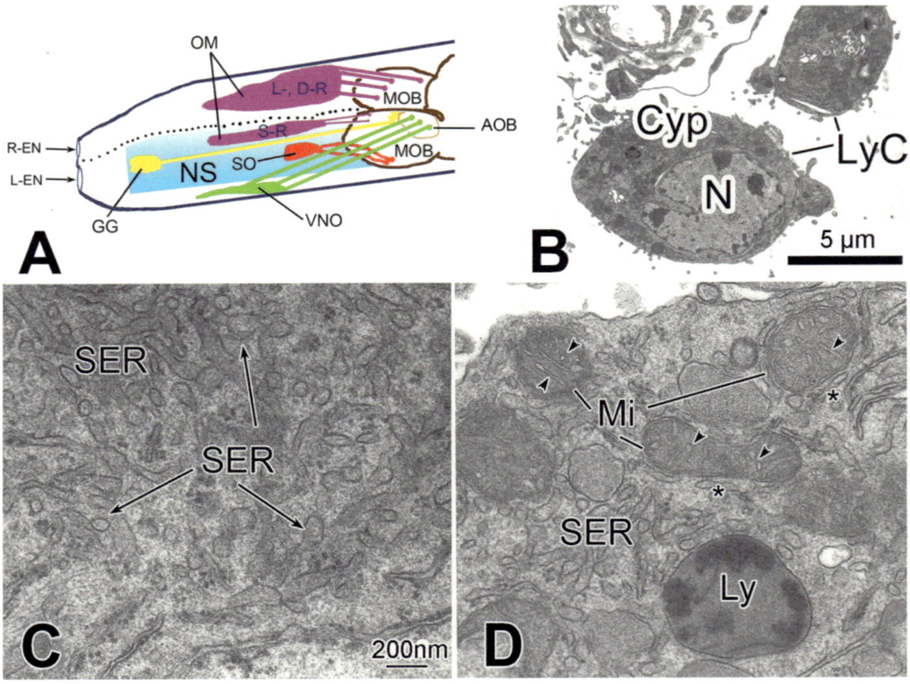

Figure 1.
*Schematic drawing showing four chemosensory nasal mucosae (**A**), transmission electron microscopic (TEM) images of a Leydig cell in the rat testis (**B–D**). **A**. Schematic drawing showing the location of Greuenberg ganglion (GG), vomeronasal organ (VNO), olfactory mucosa (OM), and septal organ (SO) and their brain centers. Axon bundles originating from the GG, OM, and SO make their terminals in the main olfactory bulb (MOB), whereas those from the VNO end in the accessory olfactory bulb. Adapted from [10] with permission (partially changed). **B**. Low magnification of a Leydig cell that contains cell nucleus (N) and cytoplasm (Cyp). **C**. The area of smooth endoplasmic reticulum (SER) in the Cyp of the Leydig cell in B. **D**. Mitochondrion (Mi) with vesicular inner structures (arrowheads) and SER. Note that single cisterna of SER (asterisk) is closely associated with the outer membrane of a Mi. Ly, lysosome.*

male olfactory investigation of volatile body odors. On the other hands, it has been reported that estradiol makes various effects on the VNO. The volume of VNO in male rats was reported to be significantly larger that of female rats. This sexual dimorphism in rat VNO is established in the perinatal days when estradiol is converted from gonadal testosterone and makes an organization effect on VSCs [15, 16]. Estradiol induced increased immunoreactivity (IR) for immediate early gene in specific population of VSCs in mice [17]. Expression of VNO receptors is sexually dimorphic and influenced by estradiol and testosterone [18].

Steroid hormones are produced and secreted by endocrine cells in the adrenal cortex [19], Leydig cells or intestinal cells of Leydig in the testis [20], granulosa cells of the ovulatory follicle and cells of the corpus luteum in the ovary [21], and brain cells [22]. One of the subcellular properties of steroid-producing cells is the presence of extremely developed smooth endoplasmic reticulum (SER, [23–25]). Images of rat Leydig cells that were obtained by transmission electron microscope (JEM1230, JEOL, Akishima-shi, Tokyo, Japan) are shown in **Figures 1–D**. In addition to tubular cisternae of SER (**Figure 1C**), swollen vesicular mitochondria are seen and associated with cisterna of SER (**Figure 1D**). In fact, Al-Amri et al. [26] demonstrated in Leydig cells in the house gecko, hemidactylus flaviviridis, that swollen vesiculated mitochondria are associated between SER. Lysosomes, which are described in the cytoplasm of Leydig cells [27], are frequently seen in the rat Leydig cells (**Figure 1D**).

All of steroid-producing cells express the cholesterol side chain cleavage enzyme (P450scc), which converts cholesterol into pregnenolone [28]. In Leydig cells of the testes, for example, pregnenolone is eventually converted into testosterone, which is the most potent androgen, via metabolic pathways containing 17β-hydroxysteroid dehydrogenases (HSDs). Based on several references [19, 20, 29–31], the main pathway of steroidogenesis from cholesterol to E2 is shown in **Figure 2**. The terminal enzyme in the sex-steroid-producing pathway is aromatase (CYP19 or P450arom). Granulosa cells of the ovulatory follicle and cells of the corpus luteum in the ovary, for instance, express aromatase (P450arom, [21]). Aromatase converts testosterone

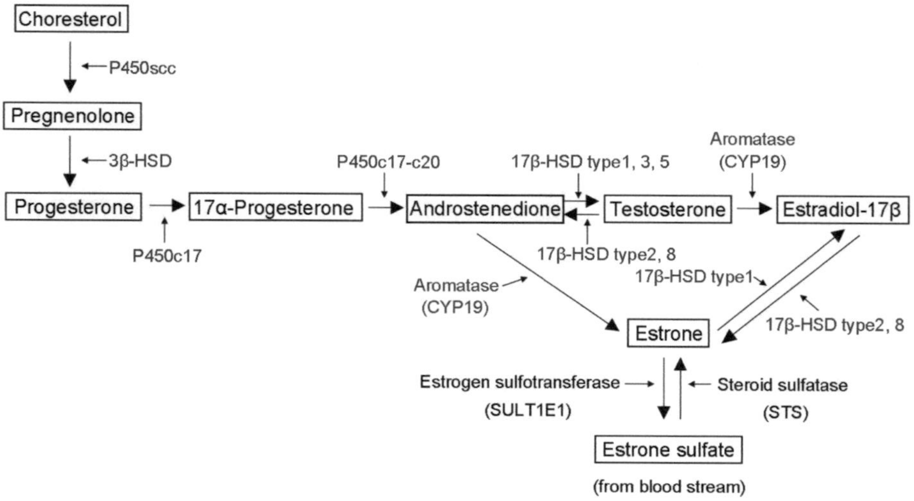

Figure 2.
Biosynthetic pathways of sex-steroids based on [20, 21, 29–31]. Enzymes involved in catalyzed changes are highlighted in blue color. P450scc, cholesterol side-chain cleavage enzyme; HSD, hydroxysteroid dehydrogenase; P450c17, steroid 17alpha-hydroxylase: CYP19, cytochrome P450 family 19.

to E2, an active type of estrogen [20]. The above enzymes are expressed by the cells in the ovary [21]. Mindnich et al. [29] introduced two major roles of 12 17β-HSDs; one, for example, 17β-HSD type 1 (17β-HSD-1) is involved in conversion of an inactivated estrogen, estrone, into an activated estrogen, estradiol, and 17β-HSD type 2 (17β-HSD-2) is involved in its opposite reaction.

Among constituent cells of the OM, olfactory sustentacular cells (SCs) surrounding OSCs are rich in SER [1–3, 32, 33]. Also, Frisch [32] reported in mice as experimental models that acinar cells (ACs) of the Bowman's glands (BGs), which are associated glands of the OM, contained large amount of SER. Before describing our findings relating to E2 in the OM, we will briefly introduce constituent cells of the rat OM, and ultrastructure of olfactory SCs and ACs.

2. Anatomy of olfactory SCs and ACs of BGs

A schematic drawing showing constituent cells of the rodent OM is shown in **Figure 3A**. The olfactory epithelium (OE) contains olfactory SCs mature and immature OSCs, globose and horizontal basal cells, and microvillar cells. Olfactory SCs and dendrites of mature OSCs are located side by side. Although duct cells of BGs penetrate into the OE and reach the epithelial surface, ACs are exclusively located in the lamina propria (LP) below the basement membrane. Using a field-emitted scanning microscope (FE-SEM, JSM6700F, JEOL), we were able to understand relationships between olfactory dendrites and SCs three-dimensionally. For example, a cylindrical dendrite between two SCs goes up to the epithelial surface to make a dendritic terminal (DT) can be seen (**Figure 3B**).

By transmission electron microscopy (TEM), cilia originating from dendritic terminals of OSCs are clearly seen (**Figure 3C** and **D**). Olfactory cilia are divided into a short proximal portion (about 400 nm at maximal width/diameter) and long distal one (100–50 nm at diameter). The proximal cilia tend to extend diagonally upward from the dendritic terminal and begin to narrow at around 1–1.5 μm and shift to the distal portion of cilia. By contrast, microvilli (about 100 nm at maximal width) extend substantially straight from the apical membranes of olfactory SCs (**Figure 3C**). At cross-sectional plate, proximal cilia extend radically from the dendritic terminals and distal cilia run approximately parallel to the epithelial surface (**Figure 3D**, [10]).

Using mice as experimental models, Makino et al. [33] classified the SER of olfactory SCs into two types, stacked lamellar and reticular-shaped SER. Using adult Sprague–Dawley rats (*Rattus norvegicus*) as experimental animals, we reported that the supranuclear region of olfactory SCs contained well-developed SER with a reticular form and myeloid bodies (MBs) [34]. The MBs are generally seen in retinal pigment epithelial cells of many species [35–39]. The MBs in olfactory SCs are composed of rows of narrow cisternae bounded by flat membranes, which were narrower than the cisternae of the SERs. They were stacked from 5 to 20 rows, ranging between 200 and 700 nm in width, and arranged mainly parallel to the lateral membrane of the olfactory SCs. The supranuclear region of the OE and SERs and MBs are shown in **Figure 4A** and **B**, respectively. Each cisterna of a MB, averaging 11 nm in width, frequently contains electron-dense material. We demonstrated that MBs and SERs are interchangeable (see **Figure 5a** of [34]). Vesicular mitochondria and associated cisternae, which are shown in Leydig cell (**Figure 2C**), are seen in the olfactory SCs as well (**Figure 4C**). In combination with the osmium-DMSO-osmium procedure [40]

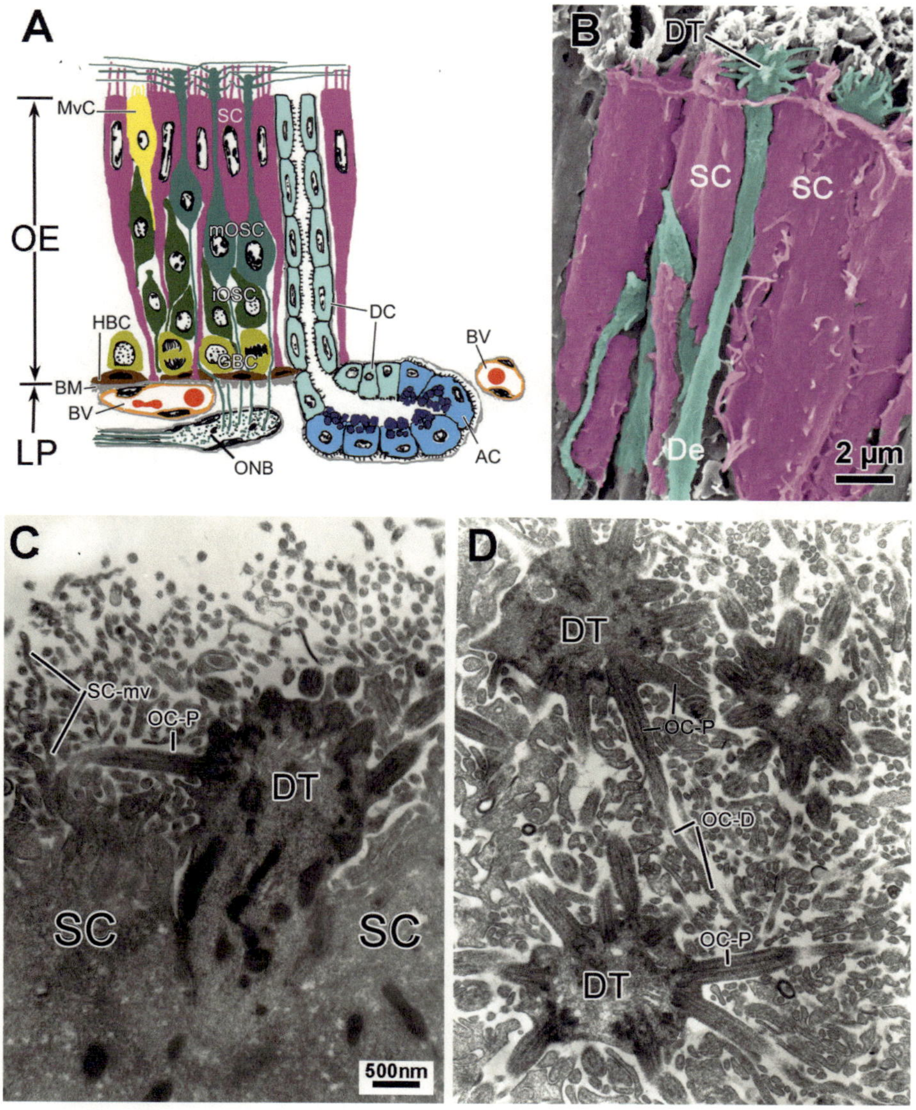

Figure 3.
Morphology of rodent olfactory mucosa (OM). ***A****. Schematic drawing of constituent cells and tissue components in the olfactory epithelium (OE) and in the lamina propria (OE) of the OM. mOSC, mature olfactory sensory cell; iOSC, immature OSC; SC, olfactory sustentacular cell; GBC, globose basal cell, HBC, horizontal basal cell; MvC, microvillar cell; BM, basement membrane: BV, blood vessel; ONB, olfactory nerve bundle; DC, duct cell of a Bowman's gland (BG); AC, acinar cell of a BG. Adapted from [10] with permission (partially changed).* ***B****. Field-emitted scanning microscopic (FE-SEM) image of an apical part of the OE, lateral view. A dendrite (De) extends upward between two olfactory sustentacular cells (SCs), reach the epithelial surface to form a terminal, dendritic terminal (Dt).* ***C, D****. TEM images of the apical part of the OE at same, with a scale bar indicated in* ***C****. adapted from reference [10] with permission (slightly changed). At longitudinal plane, proximal portion of olfactory cilia (OC-P) extending from a DT is seen, whereas microvilli (SC-mv) extend from the apical membrane of a SC (****C****). At tangential plane of the epithelial surface, OC-Ps extending radially from DTs and one of them continued to a thinner distal cilium (OC-D) are seen (****D****).*

and FE-SEM technique, we can see internal structures and relationships between MB, SER, and rough endoplasmic reticulum (RER) three-dimensionally (**Figure 4D**). Our ongoing study conducted in the OM of Suncus (*Suncus murinus*) has demonstrated the large amount of SER and characteristic mitochondria (**Figure 4E**).

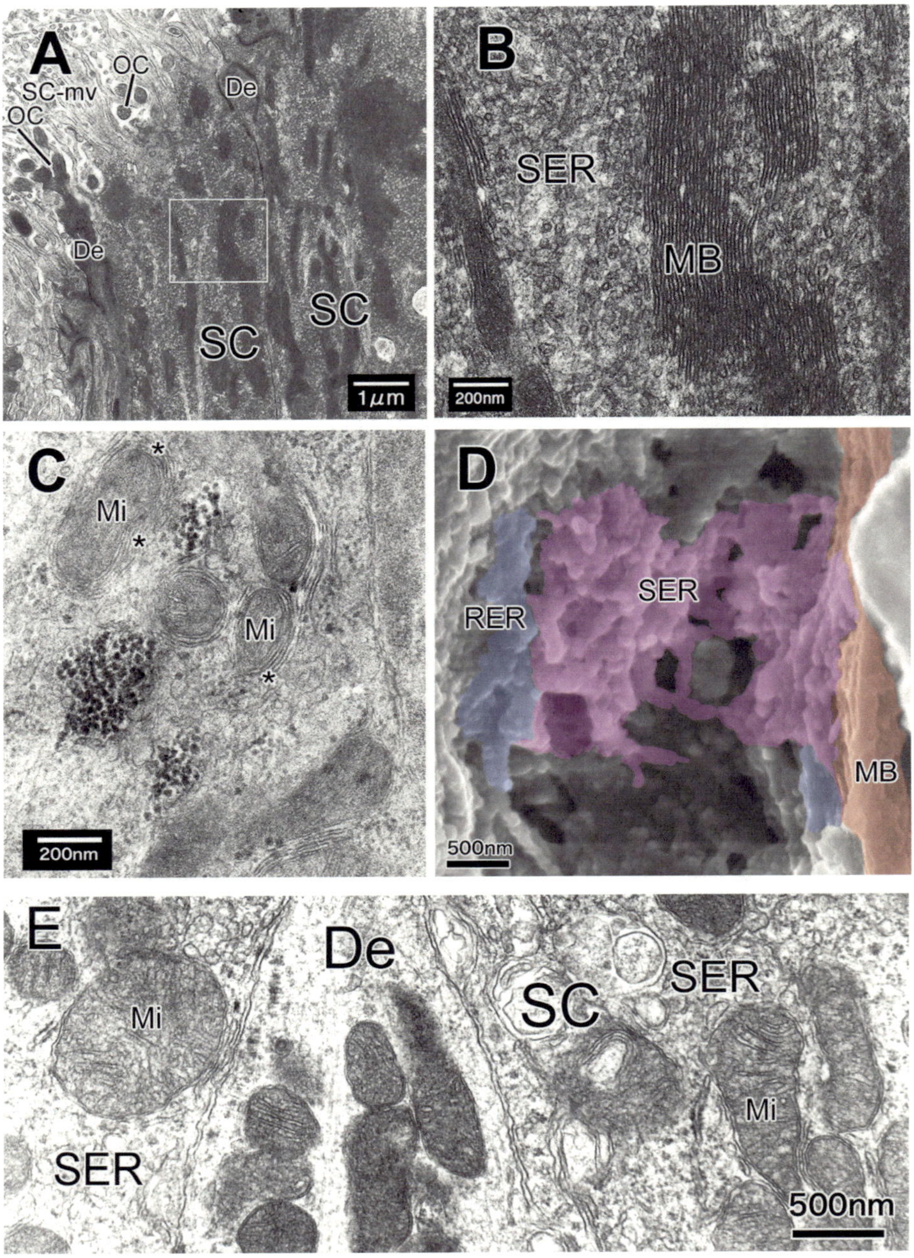

Figure 4.
TEM images (A, B, C, E) and FE-SEM image (D) of olfactory sustentacular cells. (SCs). ***A****. Low-magnified TEM image of the apical part of rat olfactory epithelium (OE), showing olfactory SCs, dendrites (De), cross sectioned olfactory cilia (OC), and microvilli of SC (SC-mv). Enlargement of a box is shown in B.* ***B****. Enlargement of the box in A, TEM image. Next to well-developed SER, myeloid bodies (MB) that characterize pigment epithelial cells in the retina are seen.* ***C****. TEM image of a rat OE showing mitochondria (Mi). Granular structures near the Mi are glycogen particles.* ***D****. FE-SEM image of partially digested cytoplasm of a rat SC. Adjacently located rough endoplasmic reticulum (RER), SER, and MB drawn by different colors are seen.* ***E****. TEM image of the Suncus OE showing two olfactory SCs and one dendrite (De) filled in-between them. Well-developed SER and swollen and vesicular Mi characterize the cytoplasm of SCs.*

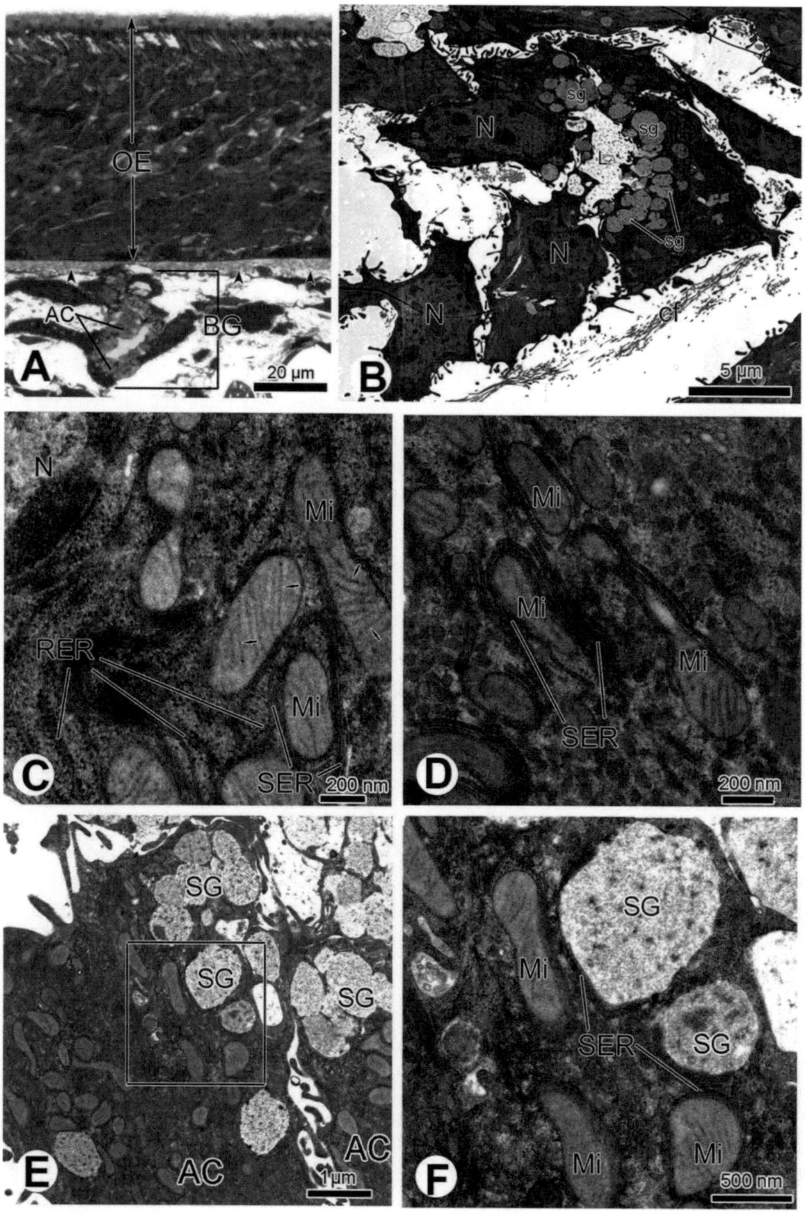

Figure 5.
*Light microscopic (A) and TEM (**B–F**) images of Bowman's glands of rats. **A**. Toluidine blue-stained semi-thin Epon-Araldite section showing olfactory epithelium (OE) and underlying lamina propria containing a Bowman's gland (BG). Acinar cells (ac) of the BG are located below the basement membrane (arrowheads) of the OE. **B**. Acinar cells that contain nuclei (N) and secretory granules (sg). Note that some of secretory granules line adjacent to the lumen (L). Collagen fibers (cf) are seen near the basal membrane of the acinar cells. **C**. The cytoplasm of a acinar cell that is close to the cell nucleus (N). Close association between the nucleus and cisternae of rough endoplasmic reticulum (RER) are seen. The cisternae of the RER appear to be continued to those of smooth endoplasmic reticulum (SER). Some of cisternae of the SER are closely associated of the outer membrane of mitochondria (Mi) that contain intenral microtubular structures (small arrows). **D**. Close association of cisternae of SER and mitochondria (Mi). Double or triple-layered cisternae of SER are attached to the outer membrane of Mi. **E**. Apical region of two acinar cells (AC) showing an aggregation of secretory granules (SG). The enlargement of the box area is shown in **F**. **F**. Close association of cisternae of SER, are even mitochondria (Mi), and secretory granules (SG). Note that cisternae of SER are even attached to the membrane of secretory granules.*

BGs (**Figure 3A**) are called attached glands of the OE and provide serous and mucous products. The secretions from these glands provide most of the mucus covering the olfactory epithelial surface, which protects the epithelial surface from drying and temperature extremes and helps to prevent damage caused by infectious agents and noninfectious particles [3]. Thus, it does make sense that BGs secrete secretory immunoglobulin A and J chain, lactoferrin, and lysozyme [41]. It has been demonstrated that secretory products of rodent BGs contain sulfated carbohydrate substances [42, 43]. By lectin histochemical technique, we demonstrated that secretory granules of rat BGs contain glycoconjugates with internal N-acetyl glucosamine (GlcNAc) residues [44].

Our ongoing study on rat BGs has indicated that ultrastructural characteristics of ACs of BGs are very similar to those in steroid-producing cells [23–26]. For example, light and transmission electron microscopic images of rat BGs are shown in **Figure 5**. ACs of a BG are located in the lamina propria below the basement membrane of OE (**Figure 5A**). Secretory granules (SGs) are seen in the supranuclear region of an AC (**Figure 5B**). The rough endoplasm reticulum (RER) that contains double- or triple-layered cisternae attached with ribosomes is easily seen in the perikaryal region and is connected with SER (**Figure 5C**). The presence of considerably well-developed RER may be a reflection of GlcNAc-containing glycoconjugates present in SGs of rat BG-ACs [44], because addition of GlcNAc residues to the backbone proteins of GlcNAc-containing glycoconjugates takes place in the RER [45, 46]. In contrast to SER in olfactory SCs, cisternae of SER of ACs appear to be filled with considerably high-dense materials (**Figure 5C** and **D**). Also, Mi containing internal microtubular structures are prominently seen (**Figure 5C**). Furthermore, inner cisterna of double- or triple-layered SER cisternae is attached to the outer membrane of Mi (**Figure 5D**). SGs are not homogenous in electron density (**Figure 5E** and **F**) and appear to be fused in the juxtaluminal region. In addition, cisternae of SER are even attached to the membrane of SGs (**Figure 5F**). Therefore, we would make a statement that we have extended Frisch's observation [32] on mouse ACs of BGs in that they contain well-developed SER and provided new ultrastructural characteristics of rat ACs.

3. Expression of P450scc, 17β-HSD-1, 17β-HSD -2, and aromatase in the rat OM

Using RT-PCR analyses, we demonstrated that P450scc, 17β-HSD-1, 17β-HSD -2, and aromatase are expressed in the nasal mucosae of both sexes of adult rats that contain the OM and respiratory mucosa [34]. These results strongly suggest that OM cells express mRNA for P450scc, 17β-HSD-1, 17β-HSD-2, and aromatase. Specific antisera against P450scc, 17β-HSD-1, and 17β-HSD-2 were used for Western blot (WB), a combination of multi-labeling immunofluorescence (IF) and confocal laser-scanning microscopic (CLSM) techniques, and immunoelectron microscopic (IEM) analyses. By WB analyses on nasal tissue specimens of both male and female adult rats, we detected single band at 49 kDa for anti-P450scc antiserum, that at 34.5 kDa for anti-17β-HSD-1, and that at ~43 kDa for anti-17β-HSD-2 antiserum 17β-HSD -2 [34]. These results strongly suggest that the OM cells produce proteins for P450scc, 17β-HSD-1, and 17β-HSD-2.

Antiserum to olfactory marker protein (OMP) equivocally immunostains mature OSCs [47, 48]. Using antiserum to OMP and one of the antisera of the above three enzymes, we determined that both olfactory SCs and cells in BGs contain IF for

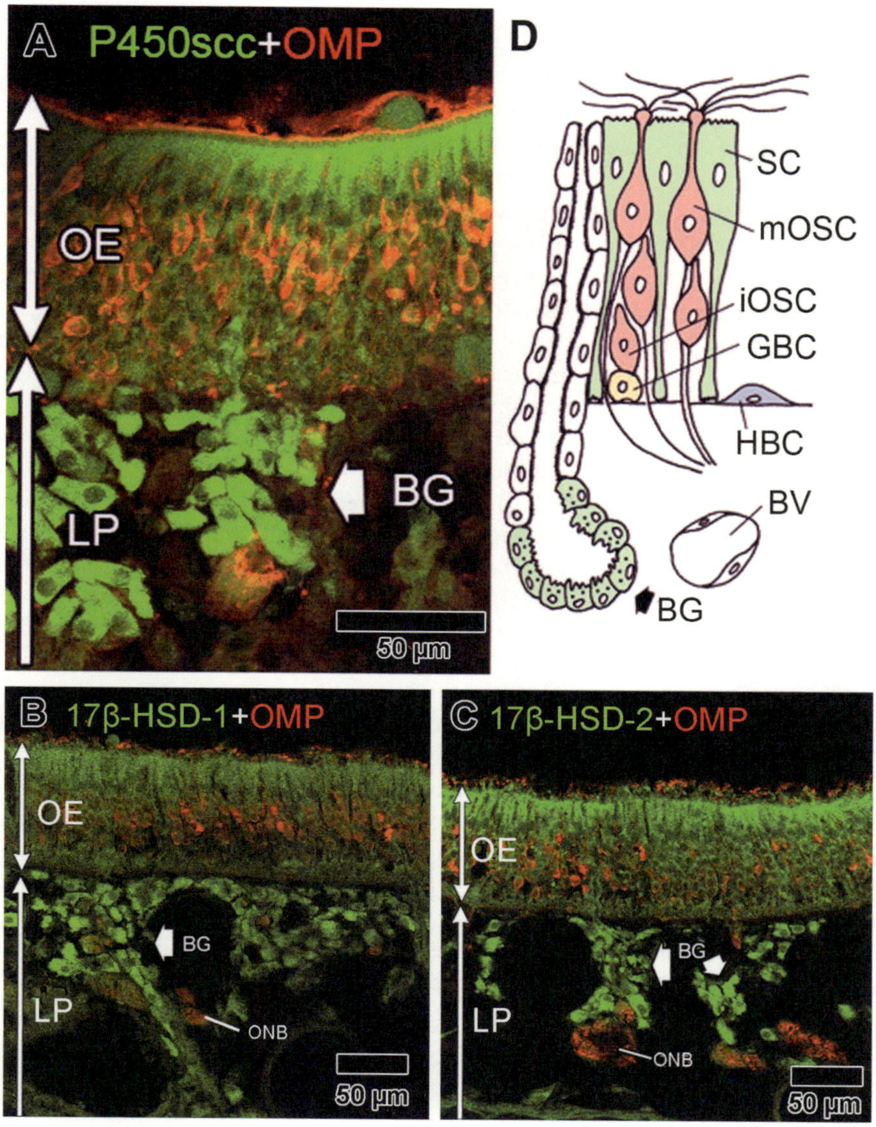

Figure 6.
Images obtained from rat OM sections by a CLSMF (A-C) and a scheme showing OM cells exhibiting immunofluorescence (IF) for P450scc, 17β-HSD-1, 17β-HSD-2, and olfactory marker protein (OMP, **D***).* **A***. Section double-labeled with antisera to OMP (red) and P450scc (green). IF for OMP is seen in somata and dendrites of mOSCs within the OE and ONB in the LP, whereas prominent IF for P450scc is seen in the apical region of the OE that is corresponding to the supranuclear cytoplasm of olfactory SCs, and cells in the BGs in the LP.* **B***. Section double-labeled with antisera to OMP (red) and 17β-HSD-1 (green). Localization of IF for OMP and 17β-HSD-1 is similar to that for OMP and P450scc in* **A***.* **C***. Section double-labeled with antisera to OMP (red) and 17β-HSD-2 (green). Localization of IF for OMP and 17β-HSD-2 is similar to that for OMP and P450scc in* **A***.* **D***. Scheme of OM in which cells containing IF for P450scc, 17β-HSD-1, and 17β-HSD-2 are indiated in green on OMP in pink. See abbreviations for the nomenclature of cell types and tissues components in* ***Figure 3A****.*

P450scc, 17β-HSD-1, and 17β-HSD -2. However, no IF was detected in mature OSCs, immature OSCs, basal cells, and olfactory nerve bundles (**Figure 6**) [34]. By IEM analyses conducted on OE cells, we demonstrated that immunoreactivity (IR) for P450scc is mainly localized in mitochondria of olfactory SCs, whereas SERs and MBs of these cells contain IR for 17β-HSD-1 and 17β-HSD-2 [34]. Similar subcellular immunolocalization of P450scc and HSDs were reported in the adrenal gland and gonads [49–51].

4. Expression of estrogen receptor β (ERβ) in the OM

The above findings strongly suggest that olfactory SCs and ACs of BGs produce E2. It is known that E2 utilizes three types of receptors, such as estrogen receptor alpha (ERα), estrogen receptor beta (ERβ), and G-protein coupled estrogen receptor 1(GPER1). Both ERα and ERβ exist as cell membrane associated form and intracellular form, whereas GPER1 exists as a transmembrane receptor [52]. Based on the results of pilot experiments, we first examined expression and cellular localization of ERβ in adult rats of both sexes. The outline of gene structure of ERβ, which is composed of eight exons, is shown in **Figure 7A**. Forward and reverse primers of 35 bases for reverse transcription-polymerase chain reaction (RT-PCR) were chosen from cDNA sequences between exon 7and 8. By RT-PCR, ERβ was expressed in nasal mucosae of both sexes as well as ovaries as a positive control (**Figure 7B**). By WB analysis using a commercially available antiserum against rabbit ERβ, ERβ protein was detected from nasal mucosae of both sexes as well as ovaries (**Figure 7C**). These results strongly suggest that OM cells in both sexes express and produce ERβ. Using antisera against OMP and ERβ, and 4′6-diamidine-2′-phenylindole dihydrochloride (DAPI) that binds to cell nucleus, triple-labeling IF technique was applied to cryostat-cut sections of OMs. Observation by a confocal laser-scanning fluorescence microscope (CLSFM, FLUOVIEW FV500-IX-UV, Olympus, Shinjuku-ku, Tokyo, Japan) indicated that mature OSCs contained very intense IF for ERβ in DTs and moderately intense IF in their supranuclear cytoplasm. Furthermore, dendrites of mature OSCs were ERβ-immunoreactive (**Figure 8A–C**). However, olfactory SCs were immune-negative for ERβ. At subcellular level, post-embedding immunogold TEM technique demonstrated that gold particles that are reflection of ERβ immunoreactivity were localized on cell membranes of both proximal and distal portions of olfactory cilia (**Figure 8D–F**). The above results [53] strongly suggest that mature OSCs, at least, contain both membranous and cytosolic ERβ.

5. E2 actions in the OM

Based on the above results, we present our working hypothesis in **Figure 9**. Since our unpublished study have demonstrated that olfactory SCs are immunoreactive for cholesterol receptor (low-density lipoprotein receptor), it is likely that the SCs take up cholesterol via capillary just beneath the basement membrane of the OE. They convert it to pregnenolone using P450scc. Since the SCs produce 17β-HSD-1, and probably aromatase as well, SCs finally produce E2 that makes biological effects on OSCs via the ERs, although the presence and cellular localization of at least two enzymes used from pregnenolone to androstenedione need to be determined (see **Figure 2**).

6. Morphological features of SER in VSCs

Presence of large amount of SER was described for the first time by Altner et al. [54]. Since then, this ultrastructural feature has been confirmed in many

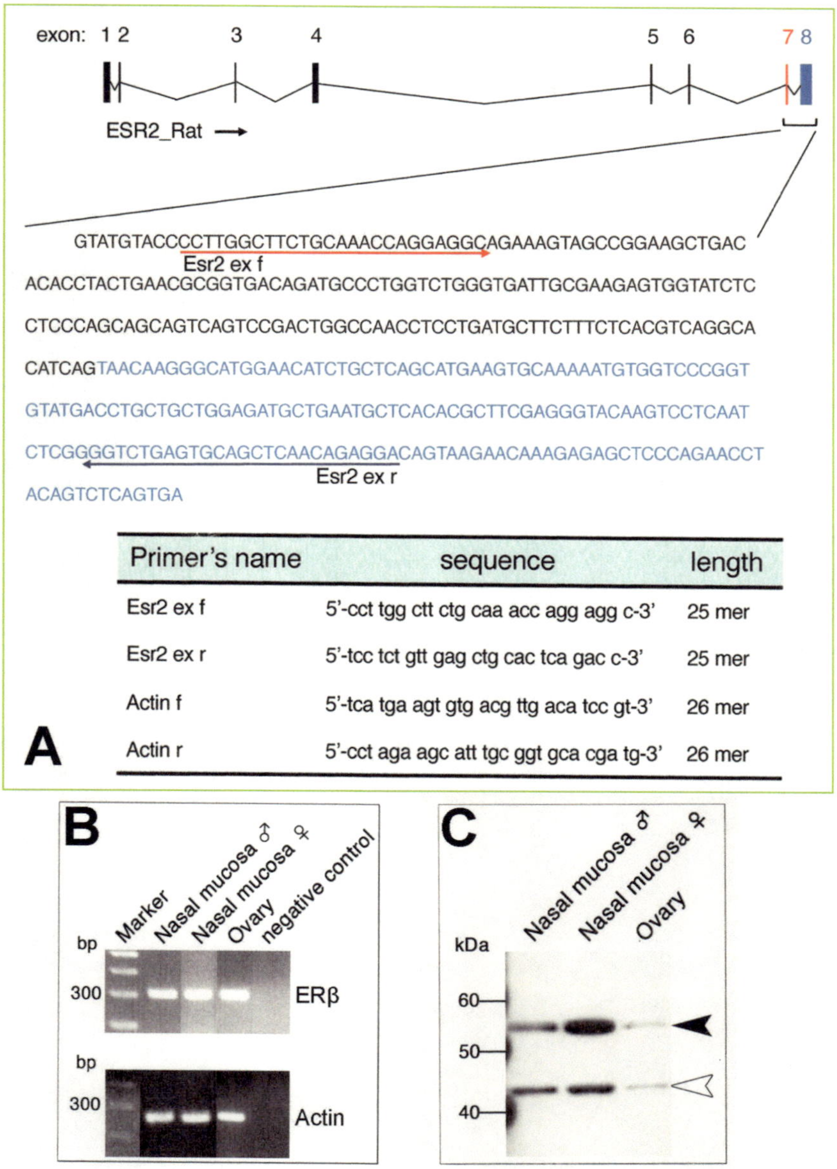

Figure 7.
A. *Brief locations of 8 exons within rat estrogen beta (ERβ) gene, and sequences of primers for RT-PCR analyses conducted in [53]. B. Expression of ERβ in nasal mucosae from males and females, and ovaries of adult rats, by a RT-PCR analysis. Note that bands with the same size are seen in nasal mucosae and ovaries as positive control tissues.* **C.** *Representative data by Western blot (WB) analysis using an antiserum to ERβ. Note that two bands detected from nasal mucosae from male and female adult rats, and rat ovaries as positive controls are the same in protein size.*

VNO-possessing animals such as amphibians [55, 56], reptiles [55, 57–60], rodents [61–64], and prosimian primates [65]. However, bioactive molecules produced by SER in VSCs had not been determined.

In each VSC soma located in the basal part of the vomeronasal sensory epithelium (VNSE) of adult rats, the area occupied by SER is very large (**Figure 10A**). Clusters of SER appear to be present in both supra- and infra-nuclear cytoplasm. An example of

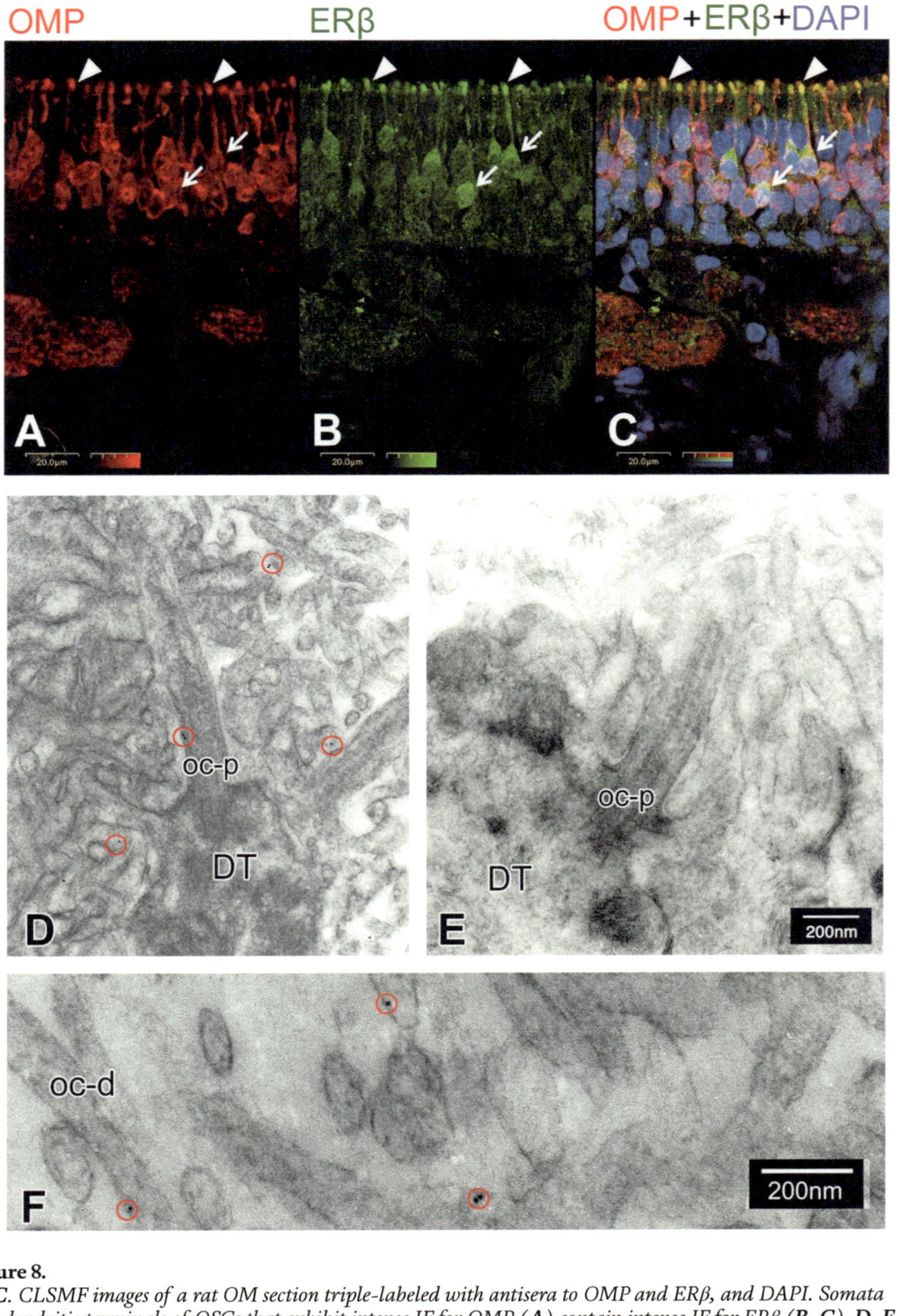

Figure 8.
*A–C. CLSMF images of a rat OM section triple-labeled with antisera to OMP and ERβ, and DAPI. Somata and dendritic terminals of OSCs that exhibit intense IF for OMP (**A**) contain intense IF for ERβ (**B**, **C**). **D–F**. TEM images of the apical part of the rat OE. **D**. Gold particles surrounded in orange, which are reflection of immunoreactivity for ERβ, are localized mainly in membranes of the proximal portion of olfactory cilia (oc-p) that are projected from a dendritic terminal (DT), by a post-embedding immunogold procedure. Gold particles surrounded in orange. **E**. A result of a negative control experiment in which the primary antiserum was omitted from staining protocol. No gold particles are found in the entire part of this image. **F**. In the distal portion of olfactory cilia (oc-d), gold particles are present in ciliary membranes and their vicinity.*

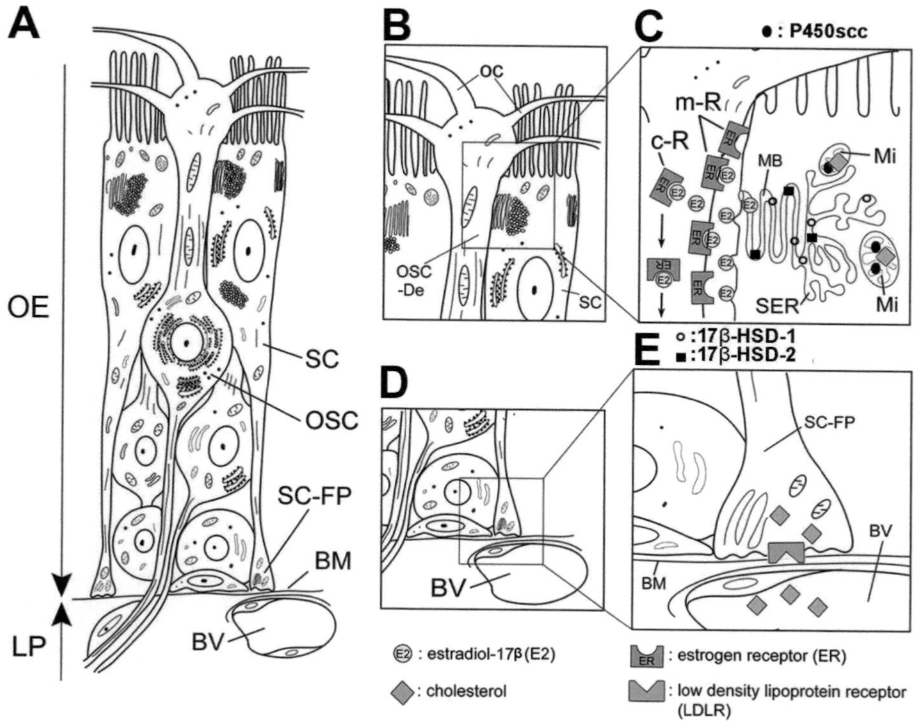

Figure 9.
Hypothetical pathway of estradiol formation and its function in the OE. ***A****. Schematic view of the olfactory mucosa showing major cellular components. Olfactory sensory cell (OSC) is located in-between sustentacular cell (SC). Foot process of SC (SC-FP) lies down the basement membrane (BM). Blood vessel (BV) is present in the LP just beneath the BV.* ***B****. Apical part of the OE. Cilia (ci) and dendrite of the ORC (ORC-De) are seen. Major molecules in the box area is schematically shown in* ***C****.* ***C****. Estardiol-17β (E2) is produced in the mitochondria (Mi), SER and MB and secreted and reacted with estrogen receptors (ER) that are located on the lateral membrane of ORC.* ***D****. Basal part of the OE. Major molecules in the box area is schematically shown in* ***E****.* ***E****. Low density lipoprotein receptors (LDLR) are circulated within the blood vessel (BV) just beneath the basement membrane (BM) and provided to the cytoplasm of the SC-FP.*

supranuclear region of a VSC soma is shown in **Figure 10B**. By TEM tomography (for details see [64]) using 200–300-thick sections, a unit volume of 700 x 700 x 150 nm of each SER area was analyzed to demonstrate microstructures of SER. Some parts of SER cisternae contained arc- and circle-like units that are complicated each other. Furthermore, some of them appeared to form tiny loops (**Figure 10B–E**).

7. Localization of 17β-HSD-1 in VSCs

By immunohistochemical techniques utilizing three antisera against 17β-HSD-1, specific staining was detected in the VNSE of adult rats (**Figure 11A** and **B**). In particular, intense immunoreactivity was seen in the perikaryal region of VSCs and in the apical region of the VNSE (**Figure 11C**). However, no specific staining was seen when the primary antiserum was omitted from the immunostaining protocol (**Figure 11D**). Using digoxigenin (Dig)-labeled cDNA antisense and sense probes for 17β-HSD-1, in

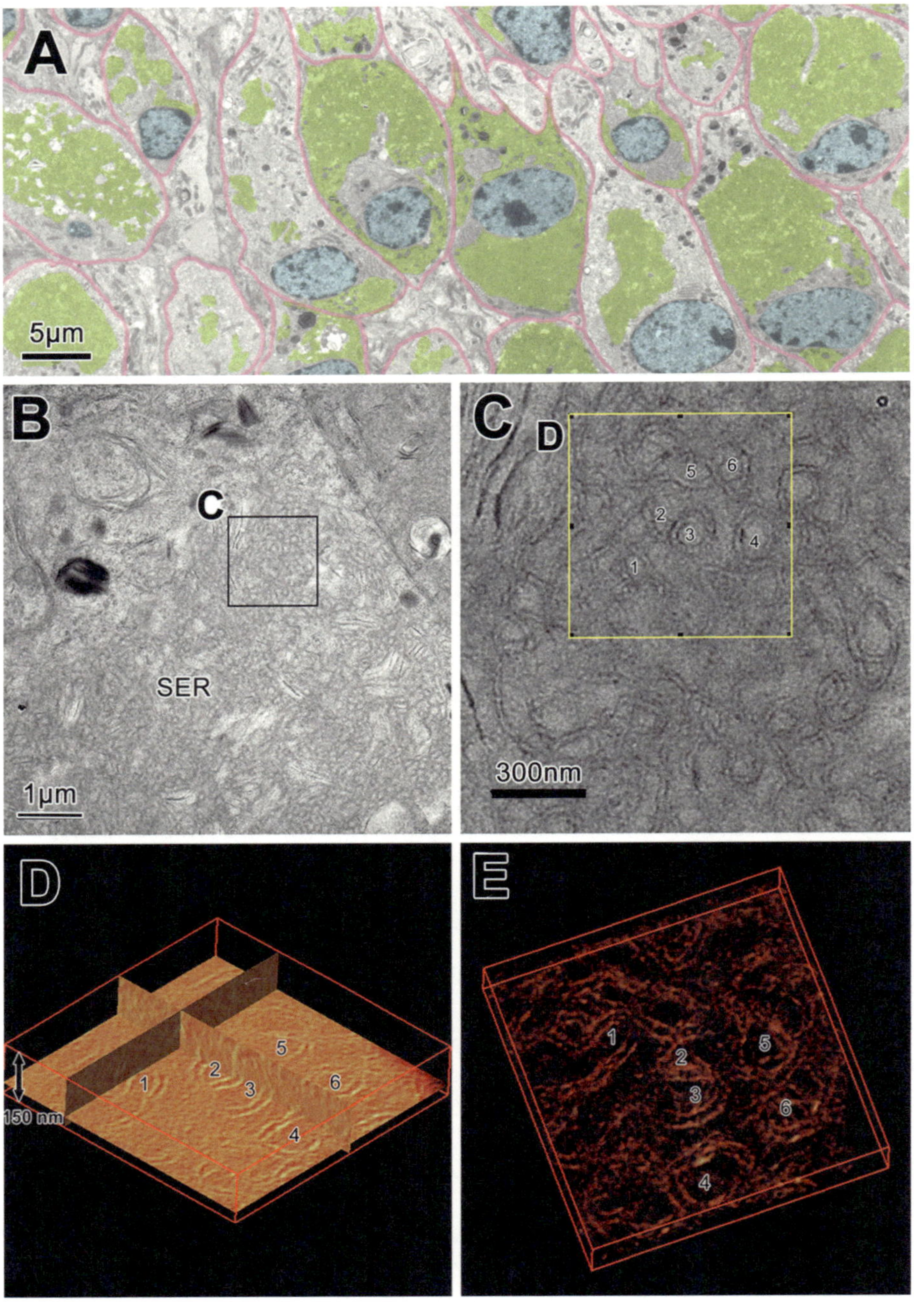

Figure 10.
*TEM images (**A–C**) of VSCs and results by TEM tomography (**D, E**). A. Low-magnified TEM image showing longitudinally-cut somata of VSCs. SERs are colored in green, nuclei in blue, and cell boundaries in magenta. Note that the cytoplasm of VSCs is largely occupied by SER. **B.** Conventional TEM image of a supranuclear region in a VSC. Enlargement of box C within the SER area is shown in **C. C.** Higher magnification of box C in panel B. Arc-and circle-like units that are numbered 1–6 in box D are clearly seen. Reconstructions of the image in box D by TEM tomography are shown in **D** and **E. D.** reconstruction of the image in box D in panel **C** by TEM tomography showing units 1–6. E. Reconstruction of image in box D in panel **C** in partially transparent mode. Complex 3D structures of units 1–6 and their looped characteristics are visible. Adapted from [64] with permission.*

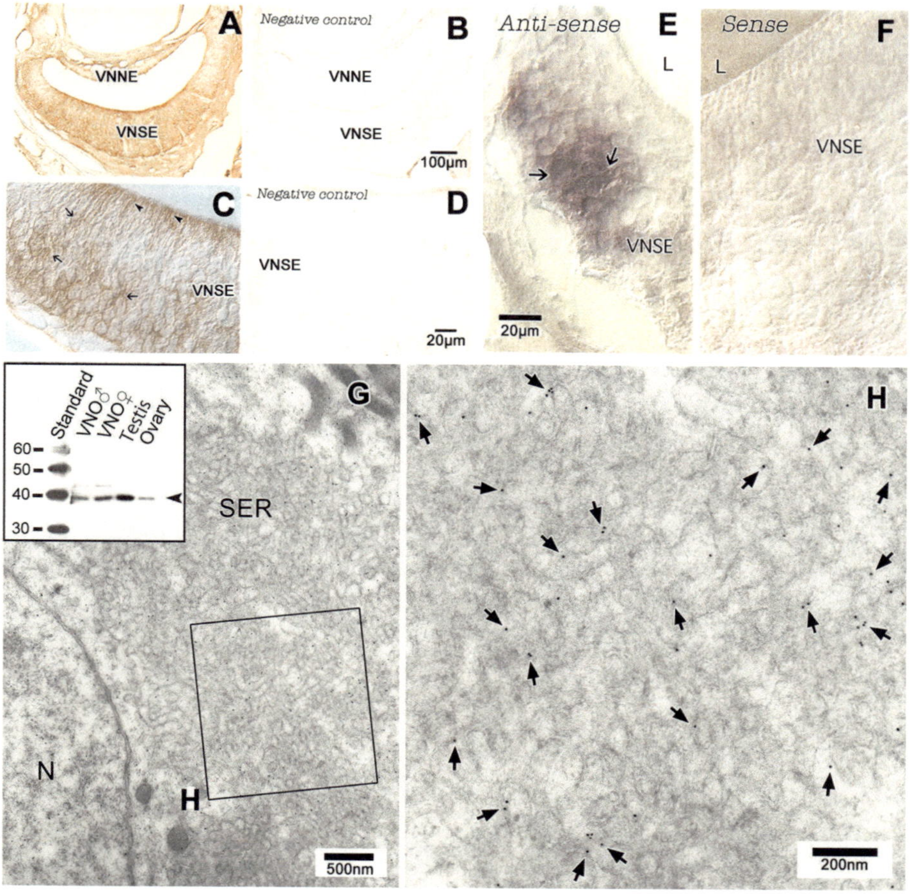

Figure 11.
Expression of 17β-HSD-1 in VNSE of adult rats. ***A–D****. Immunolocalization of 17β-HSD-1 in the VNSE and vomeronasal non-sensory epithelium (VNNE). At low magnification, intense immunoreactivity (IR) is detectable in the VNSE (A), whereas no staining is found in negative control section in which the primary antiserum was omitted from the staining protocol (B). At higher magnification, prominent IR is present in the apical part c (arrowhead) and the perikaryal region (arrow) of VSCs (C),* ***E, F****. Expression of 17β-HSD-1 in the VNSE of a rat. Using an anti-sense probe, prominent staining is seen in the perikaryal region of VSCs (E), whereas no staining is seen in a section when a sense probe was used (F).* ***G, H****. subcellular immunolocalization of 17β-HSD-1 in a VSC SER.* ***G.*** *Low magnification showing the location of SER. Boxed area is enlarged in H. N. nucleus.* ***Insert.*** *WB analysis for 17β-HSD-1 in a male VNO, female VNO, testis, and ovary. All tissues, thick bands of about 34–35 kDa are detected (indicated by an arrowhead).* ***H.*** *Higher magnification of the box in panel G showing the localization of gold particles (arrows), which reflect IR for 17β-HSD-1, localized in cisternae membranes of the SER.* ***A–F*** *are adapted from [66] and G and H from [64] with permission.*

situ hybridization (ISH) technique demonstrated that some of VSCs express 17β-HSD-1 gene (**Figure 11E** and **F**; [66]).

After confirming the presence of 17β-HSD-1 protein in the VNO by WB analyses (**Figure 11G**, insert), subcellular localization of its immunoreactivity was examined by a post-embedding immunogold TEM technique [64]. We demonstrated that gold particles, which are reflection of 17β-HSD-1 immunoreactivity, were localized in narrow cisternae of SER. Two representative images are shown in **Figure 11G** and **H**. The region shown by these images is equivalent to that analyzed by TEM tomography (**Figure 10B–E**).

In summary, it is most likely that VSCs of adult rats express 17β-HSD-1 that catalyzes conversion from androstenedione to testosterone, and that from estrone to E2. Since 17β-HSD-1immunoreactivity seems to be more intense in basally located somata of VSCs, it would be important to assess if regional difference for expression of 17β-HSD-1 is present in the VNSE.

8. VSCs express estrogen receptor α (ERα)

The VNO is known to be a pheromone-detecting sensory apparatus in rodents [5–8, 10, 67]. Hatanaka and Kiura [68] demonstrated that sensitivity of VSCs to pheromones was much more decreased in castrated mice, when compared to control mice. This data indicate that gonadal sex steroids modify physiological activities of VSCs. Setting a hypothesis that modification of sex steroids is mediated by steroid receptor, we first examined the presence of ERα in the VNO of adult rats.

Representative images from an immunohistochemical study are shown in **Figure 12**. Intense immunoreactivity for ERα is seen in the surface of the VNSE, whereas moderately intense immunoreactivity is seen in the inside of the epithelium (**Figure 12A** and **B**). Uterine glands, which are positive control tissues, contain intense immunoreactivity in cell nucleus (**Figure 12C**). It is interesting to point out

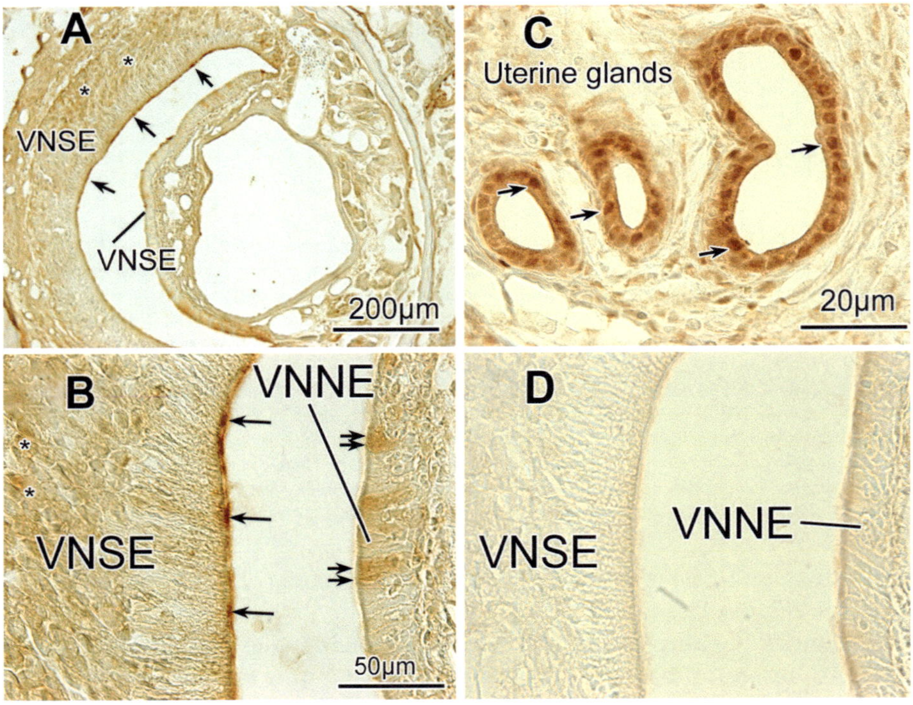

Figure 12.
Immunolocalization of ERα in the VNO (A, B, D) and uterine glands (C) as positive control. A. At lower magnification, intense IR is seen in the surface of VNSE (arrows) and its inside (asterisk). B. At higher magnification, it is clear that the surface of the VNSE contains very intensely stained loci (arrows) and less intensely stained ones. Moderated IR is also seen in somata. C. Uterine glandular cells exhibiting IR for ER. Most intense IR is seen in cell nucleus (arrows) of these cells. D. Negative control section adjacent to B. Omission of the primary antibody from the staining protocol resulted in no staining in both VNSE and VNNE.

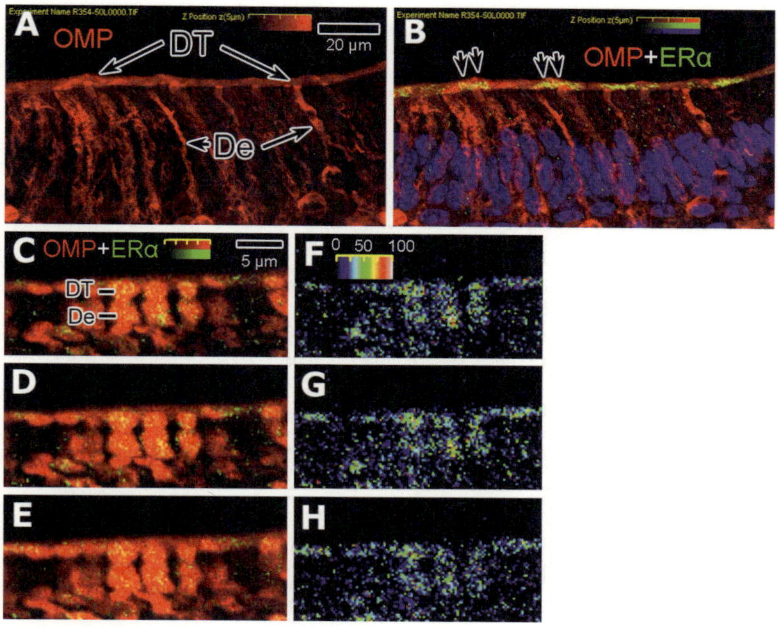

Figure 13.
CLSMF images showing colocalization of IF of OMP and ERα. A. Optical section obtained by opening a channel for OMP IF. Intense IF is seen in dendrites (De) and dendritic terminals (DT) of VSCs. B. The same section in A, obtained by opening channels for both OMP and ERα. Colocalized loci in yellow (double arrowheads) are DTs in this image. C, D, E. three serial optical sections with 0.5 µm-interval along the Z-axis, obtained by opening the channels for both OMP and ERα. Colocalized loci in yellow are continuously seen in Des and DTs. F, G, H. The same optical sections in C, D, E, obtained by opening the channel for ERα and shown in intensity-coded mode/see intensity-coded scale bar).

that immunoreactivity is also seen in the apical part of vomeronasal non-sensory epithelium (VNNE, **Figure 12B**). No specific staining is seen both in the VNSE and VNNE, when the primary antiserum was omitted from the immunostaining protocol (**Figure 12D**).

Representative CLSFM images from an IF study in which antisera against OMP and ERα, and DAPI were used are shown in **Figure 13**. Since dendrites and dendritic terminals (DTs) of VSCs exert intense IF for OMP (**Figure 13A**), it is apparent that DTs contain IF for Erα (**Figure 13B**). Optical sections containing apical dendrites show that intense IF for ERα is seen in DTs and apical dendrites just below the DTs (**Figure 13C–H**).

Representative data from a WB analysis is shown in **Figure 14**. Bands at 66 kDA that is equivalent to the size of ERα protein are detected both pellet and supernatant (sup.) specimens of VNOs and uteri that contain uterine glands.

9. Actions of sex steroids in the VNO

TEM and FE-SEM images of VSCs are shown in **Figure 15**. In contrast to OSC, many microvilli extend from the DT (**Figure 15A** and **B**). Microvilli extending from the apical membrane of vomeronasal sustentacular cells (SCs) are thicker than those of VSC (**Figure 15B**). Since intense IF for ERα was detected the apical surface of VNSE (**Figure 13**), it is most likely the both membranes and cytoplasm of microvilli.

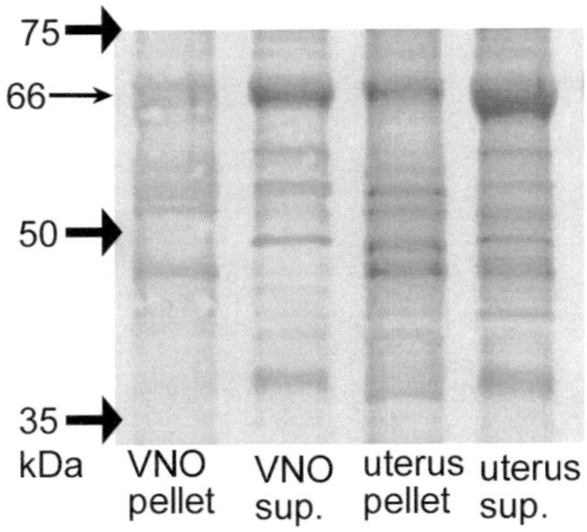

Figure 14.
Representative data by WB analysis using antiserum to ERα. A band at 66kDA, which is corresponding to molecular weight of ER protein, is detected from VNO and uterus pellet specimens that contain cell nuclei, and from their supernatants (sup.) that contain cell membranes and internal unit membrane structures, as well.

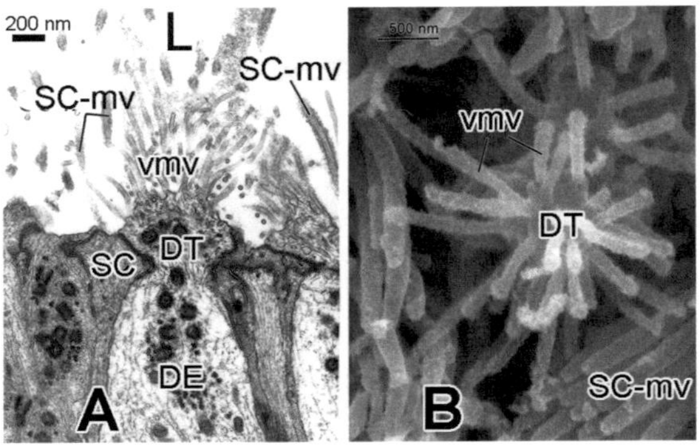

Figure 15.
A. TEM image showing the apical part of the VNSE. A dendrite (De) of a VSC reaches a lumen (L) to form a dendritic terminal (DT) from which microvilli (vmc) are extended. Adjacently located sustentacular cells (SC) have microvilli (SC-mv), as well. B. FE-SEM image of the surface of the VNSE showing DT, vm, and SC-mv. Note that short microvilli are extending radially from the DT. Adapted from [10] permission (slightly changed).

Using digoxigenin Dig-labeled cDNA antisense and sense probes for 17β-HSD-2, ISH technique demonstrated that some of VSCs express 17β-HSD-2 gene (**Figure 16** [66]). Thus, it is most likely that VSCs of adult rats produce 17β-HSD-2.

Also, we have examined immunolocalization of E2 in the rat VNSE. Antiserum to protein gene product 9.5 (PGP) is an excellent marker for both mature and immature VSCs (iVSC) [5, 69]. Using antiserum to E2, PGP, and DAPI, triple-labeling IF technique was applied to cryostat-cut sections of VNOs, and then, these sections were observed by the CLSFM. As demonstrated in **Figure 17**, intense IF for E2 is present in somata and dendrites of VSCs in a granular fashion. By contrast, moderately intense IF was diffusely seen in the supranuclear region of vomeronasal SCs. These findings would provide supporting evidence that E2 per se is present in both VSCs and SCs.

A hypothetic scheme showing origins and actions of E2 in the VNSE is presented in **Figure 18**. Since rodent VNSE contains intraepithelial capillaries [5, 62], a mature VSC (mVSC) can uptake E2 or estrone from intraepithelial capillary and give E2 to immature VSC (iVSC) via ERα. Conversion between E2 and estrone takes place in

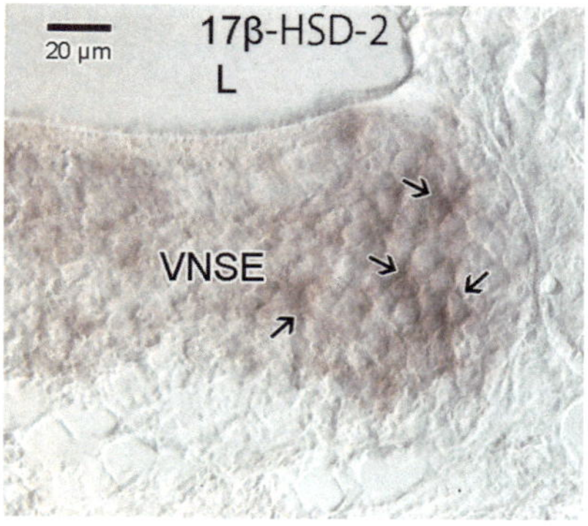

Figure 16.
Expression of 17β-HSD-2 in the VNSE of an adult rat by ISH histochemistry. VSC perikarya containing intense staining are seen in the peripheral region of the VNSE.

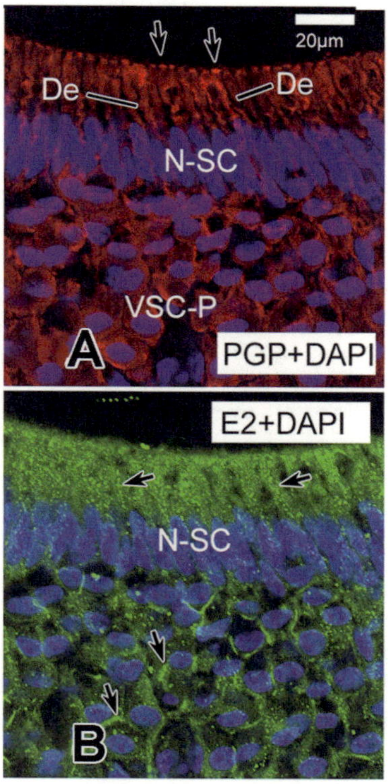

Figure 17.
CLSMF images of a section triple labeled with antisera to protein gene product 9.5 (PGP) and E2, and DAPI. A. Optical section by opening the channels for both PGP and DAPI. Intense IF for PGP are seen in the perikaryal regions of VSCs (VSC-P), dendrites (De) above the layer of sustentacular cell nuclei (N-SC), and dendritic terminals (arrows). B. Optical section by opening the channels for both E2 and DAPI. Both SCs and VSCs contains E.

SER where both 17β-HSD-1 and 17β-HSD-2 work as catalytic enzymes. It would happen that E2 from mVSC bind to growing dendritic terminal. Since brain centers of the vomeronasal system express aromatase and other steroidogenic enzymes [70], E2 may be retrogradely transported to the mVSC via neural connections.

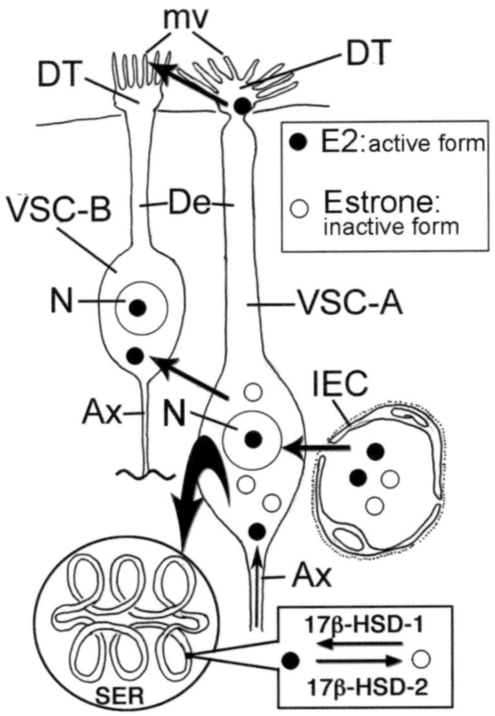

Figure 18.
Hypothetical scheme showing the metabolism of E2 in the VNSE. Since some somata of VSCs (VSC-A) are located in the close association with intraepithelial capillaries (IEC), these cells are able to take up estrone and E2 from the blood stream if they express estrogen receptors. Since VSCs express both 17β-HSD-1 and 17β-HSD-2 in their SER, they are able to convert estrone to E2, and E2 to estrone depended on physiological conditions. Also, it is likely that E2 in VSC-A is transported into other VSC, such as VSC-B at the levels of somata, and/or dendritic terminal (DT) and microvilli (mv) after transportation via dendrite (DE). It is possible that E1 is provided form via retrograde neuronal transport.

10. Conclusions

Although our studies suggest that both OM and VNO are greatly influenced by E2, relationships between cell types and enzymes/estrogen receptors are different from each other (**Table 1**) and further experiments remain to be carried out to identify cells expressing other E2-related molecules. However, at least two conclusions could be proposed here; (1) both OM and VNO are capable for producing E2, and (2) it is most likely that OSCs and VSCs contain both membrane-bound and cytosolic E2 receptors. Thus, it is most likely that cross-talk between membrane- and nuclear-initiated estrogen signaling takes place in the OM and VNO, as demonstrated in hypothalamus [71].

OSCs and VSCs are very special sensory cells, because they are bipolar neurons, but replaced by newly developed cells throughout life [3–5, 72, 73]. It is known that E2 promotes neurogenesis [74, 75], axonal [74–76], and dendritic growth [77, 78], rescues neurons [79], and promotes neuroregeneration [74, 75]. Therefore, rodent OM and VNSE are excellent experimental models for studying actions of E2, because differentiation from neuroblasts, dendritic and axonal growth, and neuronal death continuously occur in these neuroepithelia even at adult years.

Rodent OSCs share many cytochemical properties with those of human OSCs [80–84]. Thus, it is likely that E2-related enzymes and E2 receptors are expressed

	Olfactory mucosa		Vomeronasal organ		
	Sensory cell	**Sustentacular cell**	**Acinar cell**	**Sensory cell**	**Sustentacular cell**
P450scc	–	+	+	–	–
17β-HSD-1	–	+	+	+	–
17β-HSD-2	–	+	+	+	–
ERα	–	–	–	+	–
ERβ	+	–	–	–	–

Abbreviations: *P450scc, cholesterol side-chain cleavage enzyme; 17β-HSD-1, 17β-hydroxysteroid dehydrogenase type 1; 17β-HSD-2, 17β-HSD type 2; ERα, estrogen receptor alpha-type; ERβ, ER beta-type.*

Table 1.
Cellular localization of E2-related molecules in the olfactory mucosa and vomeronasal organ of adult rats.

by human OM cells, although new studies need to be initiated to examine if these molecules are present in human OM tissues. A common but striking feature in adult human OE is that patches of respiratory epithelium often appear in what was thought to be a purely olfactory region [85, 86]. Nakashima et al. [85] reported that the size of these patches varied case to case. Differential sizes of these respiratory patches may depend on differential supply of E2 on human OSCs. Parallel studies utilizing both rodent and human OMs would give us new knowledge about E2 functions, as well as new questions to be solved.

Acknowledgements

We thank Dr. Keiichi Tonosaki, former Professor of Department of Physiolosy, Meikai University, for his critical reading of this manuscript. Research projects whose results are presented here were partially supported by grants from The Promotion and Mutual Aid Corporation for Private Schools of Japan to S,T, (2008-2009 collaborative grant, and C105320018).

Author details

Shigeru Takami[1]* and Sawa Horie[2]

1 Department of Physical Therapy, Japanese School of Technology for Social Medicine, Tokyo, Japan

2 Department of Anatomy, Kawasaki Medical School, Kurashiki, Okayama, Japan

*Address all correspondence to: shigetakami@hotmail.co.jp

References

[1] Graziadei PPC. The olfactory mucosa of vertebrates. In: Beidler LM, editor. Handbook of Sensory Physiology. Vol. IV. Chemical Senses 1, Olfaction. Berlin: Springer-Verlag; 1971. pp. 27-58

[2] Graziadei PPC. The ultrastructure of vertebrates olfactory mucosa. In: Friedmann I, editor. The Ultrastructure of Sensory Ograns. Amsterdam: North Holland; 1973. pp. 267-305

[3] Farbman AI. Structure of olfactory mucous membrane. In: Barlow PW, Bray D, Green PB, Slack JMW, editors. Cell Biology of Olfaction, Developmental and Cell Biology Series 27. Cambridge, New York: Cambridge University Press; 1992. pp. 24-74

[4] Schwob JE. Neural regeneration and the peripheral olfactory system. The Anatomical Record. 2002;**269**(1):33-49. DOI: 10.1002/ar.10047

[5] Takami S. Recent progress in the neurobiology of the vomeronasal organ. Microscopy Research and Technique. 2002;**58**(3):228-250. DOI: 10.1002/jemt.10094

[6] Halpern M, Martínez-Marcos A. Structure and function of the vomeronasal system: An update. Progress in Neurobiology. 2003;**70**(3):245-318. DOI: 10.1016/s0301-0082(03)00103-5

[7] Mombaerts P. Genes and ligands for odorant, vomeronasal and taste receptors. Nature Reviews. Neuroscience. 2004;**5**(4):263-278. DOI: 10.1038/nrn1365

[8] Munger SD, Leinders-Zufall T, Zufall F. Subsystem organization of the mammalian sense of smell. Annual Review of Physiology. 2009;**71**:115-140. DOI: 10.1146/annurev.physiol.70.113006.10060

[9] Takami S, Graziadei PP. Light microscopic Golgi study of mitral/tufted cells in the accessory olfactory bulb of the adult rat. The Journal of Comparative Neurology. 1991;**311**(1):65-83. DOI: 10.1002/cne.903110106

[10] Takami S. Morphological features of nasal chemosensory mucosae of mammals —Olfactory mucosa and vomeronasal organ. Clinical Neuroscience. 2016;**34**(12):1299-1302 (Published by Chugai-Igakusha, in Japanese)

[11] Stumpf WE, Sar M. The olfactory system as a target organ of steroid hormones. In: Breipohl W, editor. Olfaction and Endocrine Regulation. London: IRL Press; 1982. pp. 11-21

[12] Woodley SK, Baum MJ. Effects of sex hormones and gender on attraction thresholds for volatile anal scent gland odors in ferrets. Hormones and Behavior. 2003;**44**(2):110-118. DOI: 10.1016/s0018-506x(03)00126-0

[13] Xiao K, Kondo Y, Sakuma Y. Sex-specific effects of gonadal steroids on conspecific odor preference in the rat. Hormones and Behavior. 2004;**46**(3):356-361. DOI: 10.1016/j.yhbeh.2004.05.008

[14] Bakker J, Honda S, Harada N, Balthazart J. Restoration of male sexual behavior by adult exogenous estrogens in male aromatase knockout. Hormones and Behavior. 2004;**46**(1):1-10. DOI: 10.1016/j.yhbeh.2004.02.003

[15] Segovia S, Guillamón A. Effects of sex steroids on the development of

the vomeronasal organ in the rat. Brain Research. 1982;**281**(2):209-212. DOI: 10.1016/0165-3806(82)90160-2

[16] Segovia S, Paniagua R, Nistal M, Guillamon A. Effects of postpuberal gonadectomy on the neurosensorial epithelium of the vomeronasal organ in the rat. Brain Research. 1984;**316**(2):289-291. DOI: 10.1016/0165-3806(84)90315-8

[17] Halem HA, Baum MJ, Cherry JA. Sex difference and steroid modulation of pheromone-induced immediate early genes in the two zones of the mouse accessory olfactory system. The Journal of Neuroscience. 2001;**21**(7):2474-2480. DOI: 10.1523/JNEUROSCI.21-07-02474.2001

[18] Alekseyenko OV, Baum MJ, Cherry JA. Sex and gonadal steroid modulation of pheromone receptor gene expression in the mouse vomeronasal organ. Neuroscience. 2006;**140**(4):1349-1357. DOI: 10.1016/j.neuroscience.2006.03.001

[19] Goodman HM. Adrenal glands. In: Basic Medical Endocrinology. 4th ed. Amsterdam: Academic Press; 2009a. pp. 61-99

[20] Goodman HM. Hormonal control of reproducttion in the male. In: Basic Medical Endocrinology. 4th ed. Amsterdam: Academic Press; 2009b. pp. 239-256

[21] Goodman HM. Hormonal control of reproducttion in the female: The menstrual cycle. In: Basic Medical Endocrinology. 4th ed. Amsterdam: Academic Press; 2009c. pp. 257-275

[22] Do Rego JL, Seong JY, Burel D, Leprince J, Luu-The V, Tsutsui K, et al. Neurosteroid biosynthesis: Enzymatic pathways and neuroendocrine regulation by neurotransmitters and neuropeptides. Frontiers in Neuroendocrinology. 2009;**30**(3):259-301. DOI: 10.1016/j.yfrne.2009.05.006

[23] Fawcett DW. Smooth endoplasmic reticulum. In: The Cell. Second ed. Philadelphia, London, Toronto: W. B Sannders Co.; 1981a. pp. 344-345 pp. 346-349, pp. 360-361

[24] Dym M. The male reproductive system. In: Weiss L, editor. Histology, Cell and Tissue Biology. 5th ed. New York: Elsevier; 1983. pp. 1000-1053

[25] Long JA. The adrenal gland. In: Weiss L, editor. Histology. Cell and Tissue Biology. 5th ed. New York: Elsevier Science Pub. Co. Inc.; 1983. pp. 1116-1133

[26] Al-Amri IS, Mahmoud IY, Waring CP, Alkindi AY, Khan T, Bakheit C, et al. The reproductive cycle of the male house gecko, Hemidactylus flaviviridis, in relation to plasma steroid concentrations, progesterone receptors, and steroidogenic ultrastructural features, in Oman. General and Comparative Endocrinology. 2013;**187**:23-31. DOI: 10.1016/j.ygcen.2013.02.040

[27] Fawcett DW. Peroxisomes. In: The Cell. Second ed. Philadelphia, London, Toronto: W. B Sannders Co; 1981b. pp. 520-521

[28] Miller WL, Bose HS. Early steps in steroidogenesis: Intracellular cholesterol trafficking. Journal of Lipid Research. 2011;**52**(12):2111-2135. DOI: 10.1194/jlr.R016675

[29] Mindnich R, Möller G, Adamski J. The role of 17 beta-hydroxysteroid dehydrogenases. Molecular and Cellular Endocrinology. 2004;**218**(1-2):7-20. DOI: 10.1016/j.mce.2003.12.006

[30] Sasano H, Suzuki T, Miki Y, Moriya T. Intracrinology of estrogens

and androgens in breast carcinoma. The Journal of Steroid Biochemistry and Molecular Biology. 2008;**108**(3-5):181-185. DOI: 10.1016/j.jsbmb.2007.09.012

[31] Miki Y, Suzuki T, Sasano H. Controversies of aromatase localization in human breast cancer—Stromal versus parenchymal cells. The Journal of Steroid Biochemistry and Molecular Biology. 2007;**106**:97-101. DOI: 10.1016/j.jsbmb.2007.05.007

[32] Frisch D. Ultrastructure of mouse olfactory mucosa. The American Journal of Anatomy. 1967;**121**(1):87-120. DOI: 10.1002/aja.1001210107

[33] Makino N, Ookawara S, Katoh K, Ohta Y, Ichikawa M, Ichimura K. The morphological change of supporting cells in the olfactory epithelium after bulbectomy. Chemical Senses. 2009;**34**(2):171-179. DOI: 10.1093/chemse/bjn074

[34] Horie S, Yamaki A, Takami S. Presence of sex steroid-metabolizing enzymes in the olfactory mucosa of rats. The Anatomical Record. 2017;**300**(2):402-414. DOI: 10.1002/ar.23497

[35] Yamada E. The fine structure of the pigment epithelium in the turtle eye. In: Smeiser GK, editor. Proceedings of the Symposium Held April 11-13, during the Seventh International Congress of Anatomists, New York, New York. New York, London: Academic Press; 1960. pp. 73-84

[36] Porter KR, Yamada E. Studies on the endoplasmic reticulum. V. Its form and differentiation in pigment epithelial cells of the frog retina. The Journal of Biophysical and Biochemical Cytology. 1960;**8**:181-205. DOI: 10.1083/jcb.8.1.181

[37] Kuwabara T. Cytologic changes of the retina and pigment epithelium during hibernation. Investigative Ophthalmology. 1975;**14**(6):457-467 PMID: 166050

[38] Braekevelt CR. Fine structure of the retinal epithelium (RPE) of the emu (Dromaius novaehollandiae). Tissue & Cell. 1998;**30**(2):149-156. DOI: 10.1016/s0040-8166(98)80063-3

[39] Braekevelt CR, Smith SA, Smith BJ. Fine structure of the retinal pigment epithelium of Oreochromis niloticus L. (Cichlidae; Teleostei) in light- and-dark adaptation. The Anatomical Record. 1998;**252**(3):444-452. DOI: 10.1002/(SICI)1097-0185(199811)252:3<444::AID-AR12>3.0.CO;2-4

[40] Ogata T, Yamasaki Y. High-resolution scanning electron microscopic studies on the tree-dimensional structure of the transverse-axial tubular system, sarcoplasmic reticulum and intercalated disc of the rat myocardium. The Anatomical Record. 1990;**228**(3):277-287. DOI: 10.1002/ar.10922280307

[41] Getchell TV, Margolis FL, Getchell ML. Perireceptor and receptor events in vertebrate olfaction. Progress in Neurobiology. 1984;**23**(4):317-345. DOI: 10.1016/0301-0082(84)90008-x

[42] Cuschieri A, Bannister LH. Some histochemical observations on the mucosubstances of the nasal glands of the mouse. The Histochemical Journal. 1974;**6**(5):543-558. DOI: 10.1007/BF01003270

[43] Kondo K, Watanabe K, Sakamoto T, Suzukawa K, Nibu K, Kaga K, et al. Distribution and severity of spontaneous lesions in the neuroepithelium and Bowman's glands in mouse olfactory mucosa: Age-related progression. Cell and Tissue Research. 2009;**335**(3):489-503. DOI: 10.1007/s00441-008-0739-9

[44] Takami S, Getchell ML, Getchell TV. Lectin histochemical localization of galactose, N-acetylgalactosamine, and N-acetylglucosamine in glycoconjugates of the rat vomeronasal organ, with comparison to the olfactory and septal mucosae. Cell and Tissue Research. 1994;**277**(2):211-230. DOI: 10.1007/BF00327769

[45] Ogawa M, Sawaguchi S, Furukawa K, Okajima T. N-acetylglucosamine modification in the lumen of the endoplasmic reticulum. Biochimica et Biophysica Acta. 2015;**1850**(6):1319-1324. DOI: 10.1016/j.bbagen.2015.03.003

[46] Zhang J, Wu J, Liu L, Li J. The crucial role of Demannosylating asparagine-linked Glycans in ERADicating misfolded glycoproteins in the endoplasmic reticulum. Frontiers in Plant Science. 2020;**11**:625033. DOI: 10.3389/fpls.2020.625033

[47] Farbman AI, Margolis FL. Olfactory marker protein during ontogeny: Immunohistochemical localization. Developmental Biology. 1980;**74**(1):205-215. DOI: 10.1016/0012-1606(80)90062-7

[48] Monti Graziadei GA, Stanley RS, Graziadei PP. The olfactory marker protein in the olfactory system of the mouse during development. Neuroscience. 1980;**5**(7):1239-1252. DOI: 10.1016/0306-4522(80)90197-9

[49] Hanukoglu I, Suh BS, Himmelhoch S, Amsterdam A. Induction and mitochondrial localization of cytochrome P450scc system enzymes in normal and transformed ovarian granulosa cells. The Journal of Cell Biology. 1990;**111**(4):1373-1381. DOI: 10.1083/jcb.111.4.1373

[50] Pelletier G, Li S, Luu-The V, Tremblay Y, Bélanger A, Labrie F. Immunoelectron microscopic localization of three key steroidogenic enzymes (cytochrome P450(scc), 3 beta-hydroxysteroid dehydrogenase and cytochrome P450(c17)) in rat adrenal cortex and gonads. The Journal of Endocrinology. 2001;**171**(2):373-383. DOI: 10.1677/joe.0.1710373

[51] Miller WL. Steroid hormone synthesis in mitochondria. Molecular and Cellular Endocrinology. 2013;**379**(1-2):62-73. DOI: 10.1016/j.mce.2013.04.014

[52] Pillerová M, Borbélyová V, Hodosy J, Riljak V, Renczés E, Frick KM, et al. On the role of sex steroids in biological functions by classical and non-classical pathways. An update. Frontiers in Neuroendocrinology. 2021;**62**:100926. DOI: 10.1016/j.yfrne.2021.100926

[53] Takami S, Tateno K, Horie S, Hasegawa R. Immunolocalization of an estrogen receptor in the rat olfactory mucosa. In: The 34th Kanto-shibu's Annual Branch Meeting of Jap Soc of Microscopy. Poster Presentation (in Japanese). Tokyo, Japan: The Japanese Society of Microscopy; 2010

[54] Altner H, Müler W, Brachner I. The ultrastructure of the vomero-nasal organ in Reptilia. Zeitschrift für Zellforschung und Mikroskopische Anatomie. 1970;**105**(1):107-122. DOI: 10.1007/BF00340567

[55] Kolnberger I. Vergleichende Untersuchungen am Rüechepitel, insbesondere des Jacobsonshen Organs von Amphibien, Reptilien und Säugetieren. Zeitschrift für Zellforschung und Mikroskopische Anatomie. 1971;**122**(1):53-67 PMID: 4330717

[56] Franceschini F, Sharbati A, Zancanaro C. The vomeronasal organ in the frog, Rana esculenta. An electron microscopy study. Journal of Submicroscopic Cytology and

Pathology. 1991;**23**(2):221-231 PMID: 2070348

[57] Graziadei PPC, Tucker D. Vomeronasal receptors in turtles. Zeitschrift für Zellforschung und Mikroskopische Anatomie. 1970;**105**(4):498-514. DOI: 10.1007/BF00335424

[58] Kratzing J. The fine structure of the olfactory and vomeronasal organ of a lizard (Tiliqua scincoides scincoides). Cell and Tissue Research. 1975;**156**(2):239-252. DOI: 10.1007/BF00221807

[59] Wang RT, Halpern M. Light and electron microscopic observations on the normal structure of the vomeronasal organ of garter snakes. Journal of Morphology. 1980;**164**(1):47-67. DOI: 10.1002/jmor.1051640105

[60] Takami S, Hirosawa K. Electron microscopic observations on the vomeronasal sensory epithelium of a crotaline snake, Trimeresurus flavoviridis. Journal of Morphology. 1990;**205**(1):45-61. DOI: 10.1002/jmor.1052050106

[61] Ciges M, Lavella T, Gayaso M, Sanchez G. Ultrastructure of the organ of Jacobson and comparative study with olfactory mucosa. Acta Oto-Laryngologica. 1977;**83**(1-2):47-58. DOI: 10.3109/00016487709128812

[62] Breipohl W, Bhatnager KP, Blank M, Mendoza AS. Intraepithelial blood vessels in the vomeronasal neuroepithelium of the rat. Cell and Tissue Research. 1981;**215**(3):465-473. DOI: 10.1007/BF00233523

[63] Taniguchi K, Mocnhizuki K. Morphological studies on the vomeronasal organ in the golden hamster. Nihon Juigaku Zasshi. 1982;**44**(3):419-426. DOI: 10.1292/jvms1939.44.419

[64] Takami S, Hasegawa R, Shima Y, Horie S, Sato Y, Koyama S, et al. Analyses of smooth endoplasmic reticulum in vomeronasal receptor cells. Journal of the Kyorin Medical Society. 2009;**40**(3):24-33

[65] Loo SK, Kanagasuntheram R. The vomeronasal organ in tree shrew and slow loris. Journal of Anatomy. 1972;**112**(2):165-172 PMID: 5077189. PMCID: PMC1271190

[66] Akamine Y. Study on distribution of 17β-hydroxysteroid dehydrogenases in the vomeronasal organ. In: [Bachelor's Thesis Written in Japanese] Department of Medical Technology. Tokyo, Japan: Kyorin University Faculty of Health Sciences; 2006

[67] Takami S, Hasegawa R, Nishiyama F. A study for the distribution of an estrogen receptor in the vomeronasal organ. The Japanese Journal of Taste and Smell Research. 2006;**13**(3):575-578

[68] Hatanaka T, Kiura H. Effects of sex steroid hormones on the sensitivity of vomeronasal receptor neurons in mice. In: 14th Inter Symp on Olfaction and Taste/ 38th Jap Assoc for Study of Taste and Smell. Osaka, Japan: The Japanese Association of the Study of Taste and Smell; 2004 Abstracts 83

[69] Takami S, Getchell ML, Yamagishi M, Albers KM, Getchell TV. Enhanced extrinsic innervation of nasal and oral chemosensory mucosae in keratin 14-NGF transgenic mice. Cell and Tissue Research. 1995;**282**(3):481-491. DOI: 10.1007/BF00318880

[70] Do Rego JL, Seong JY, Burel D, Leprince J, Luu-The V, Tsutsui K, et al. Neurosteroid biosynthesis: Enzymatic pathways and neuroendcrine regulation by neurotransmitters and neuropeptides.

Frontiers in Neuroendocrinology. 2009;**30**(3):259-301. DOI: 10.1016/j.yfrne.2009.05.006

[71] Roepke AT, Qiu J, Bosch MA, Rønnekleiv OK, Kelly MJ. Cross-talk between membrane-initiated and nuclear-initiated oestrogen signalling in the hypothalamus. Journal of Neuroendocrinology. 2009;**21**(4):263-270. DOI: 10.1111/j.1365-2826.2009.01846.x

[72] Graziadei PPC. Functional anatomy of the mammalian chemoreceptor system. In: Müler-Schwarze D, Mozell M, editors. Chemical Signals in Vertebrates. New York: Plenum Press; 1977. pp. 435-454

[73] Graziadei PPC. Olfactory development. In: Coleman JR, editor. Development of Sensory Systems in Mammals. New York: John Wiley & Sons, Inc; 1990. pp. 519-566

[74] Azcoitia I, Arevalo MA, Garcia-Segura LM. Neural-derived estradiol regulates brain plasticity. Journal of Chemical Neuroanatomy. 2018;**89**:53-59. DOI: 10.1016/j.jchemneu.2017.04.004

[75] Azcoitia I, Barreto GE, Garcia-Segura LM. Melecular mechanisms and cellular events involved in the neuroprotective actions of estradiol. Analysis of sex differences. Frontiers in Neuroendocrinology. 2019;**55**:100787. DOI: 10.1016/j.yfrne.2019.100787.Epub 2019 Sep 9

[76] Carrer HF, Cambiasso MJ. Brito B Gorosito S. Annals of the New York Academy of Sciences. 2003;**1007**:306-316. DOI: 10.1196/annals.1286.029

[77] Tsutsui K. Neurosteroids in the Purkinje cell: Biosynthesis, mode of action and functional significance. Molecular Neurobiology. 2008;**37**(2-3):116-125. DOI: 10.1007/s12035-008-8024-1.Epub 2008 Jun 3

[78] Haraguchi S, Sasahara K, Shikimi H, Honda S, Harad N, Tsutui K. Estradiol promotes purkinje dendritic growth, spinogenesis, and synaptogenesis during neonatal life by inducing the expression of BDNF. Cerebellum. 2012;**11**(2):416-417. DOI: 10.1007/s12311-011-0342-6

[79] Lebesgue D, Chevaleyre V, Zukkin RS, Etgen AM. Estradiol rescues neurons from global ischemia-induced cell death: Multiple cellular pathways of neuroprotection. Steroids. 2009;**74**(7):555-561. DOI: 10.1016/j.steroids.2009.01.003.Epub 2009 Jan 20

[80] Chen Y, Getchell ML, Ding X, Getchell TV. Immunolocalization of two cytochrome P450 isozymes in rat nasal chemosensory tissue. Neuroreport. 1992;**3**(9):749-752. DOI:10.1097/00001756-199209000-00007

[81] Takami S, Getchell ML, Chen Y, Monti-Bloch L, Berliner DL, Stensaas LJ, et al. Vomeronasal epithelial cells of the adult human express neuron-specific molecules. Neuroreport. 1993;**4**(4):375-378. DOI:10.1097/00001756-199304000-00008

[82] Yamagishi M, Takami S, Getchell TV. Ontogenetic expression of spot 35 protein (calbindin-D28k) in human olfactory receptor neurons and its decrease in Alzheimer's disease patients. The Annals of Otology, Rhinology, and Laryngology. 1996;**105**(2):132-139. DOI: 10.1177/000348949610500208

[83] Yamagishi M, Getchell ML, Takami S, Getchell TV. Increased density of olfactory receptor neurons immunoreactive for apolipoprotein E in patients with Alzheimer's disease. The Annals of Otology, Rhinology, and

Laryngology. 1998;**107**(5 Pt 1):421-426. DOI: 10.1177/000348949810700511

[84] Nibu K, Li G, Zhang X, Rawson NE, Restrepo D, Kaga K, et al. Olfactory neuron-specific expression of NeuroD in mouse and human nasal mucosa. Cell and Tissue Research. 1999;**298**(3): 405-414. DOI: 10.1007/s004419900098

[85] Nakashima T, Kimmelman CP, Snow JB Jr. Structure of human fetal and adult olfactory neuroepithelium. Archives of Otolaryngology. 1984;**110**(10):641-646. DOI: 10.1001/archotol.1984.00800360013003

[86] Morrison EE, Costanzo RM. Morphology of olfactory epithelium in humans and other vertebrates. Microscopy Research and Technique. 1992;**23**(1):49-61. DOI: 10.1002/jemt.1070230105

Section 3

Estrogen and the Prostate

Chapter 5

Permissive Role of Estrogens in Prostate Diseases

José Locia Espinoza and Luz Irene Pascual Mathey

Abstract

Estrogens are steroid hormones that act through their receptors (ER) α and β. They are involved mainly in female physiology but, in males, influence the homeostasis of some tissues like the prostate. In this organ, estrogens promote or limit cell proliferation depending on the activated receptor, with implications for the pathophysiology of benign prostatic hyperplasia (BPH) and prostate cancer (PCa). ERα promotes proliferation while ER β is a protective factor against proliferative diseases. However, in the advanced stages of PCa, ERβ has a permissive role in prostate cells, increasing patient mortality. These effects are mediated by activating androgen-independent signaling pathways that promote proliferation. Another essential aspect of ER actions is the regulation of its expression. Steroid hormones participate in this process, but some non-steroid factors, like environment and epigenetic marks, influence the prostate's physiopathology. Knowledge of these and other aspects of estrogenic action in the prostate will contribute to developing strategies for treating and preventing BPH and PCa. For this reason, this chapter will review the main aspects of estrogens' permissive role in prostate diseases.

Keywords: Estrogen receptors, permissive role, prostate physiology

1. Introduction

Estrogens are steroid hormones that have a very well-characterized morphogenic role since they participate in the development of the ovary, uterus, mammary gland, prostate, lungs, and brain, among others [1]. The primary estrogen identified is Estradiol (E2), to which most of the actions of estrogens are attributed, which act through binding to specific receptors, the ER [2]. It has been shown that the diverse functions in which estrogens intervene depend on the adequate balance in the signaling of their ER (ERα and ERβ). Although the actions of estrogens have been well characterized in female physiology, it has a central role in the maintenance of male sexual function and the development of prostate diseases [3]. Concerning the above, estrogenic stimulation can cause the activation of different intracellular signaling pathways that can influence the development of the prostate gland both in the embryonic stage and in adulthood and intervene in the development of pathologies [2, 4]. Therefore, this review will address the role of estrogens and their receptors in prostate pathophysiology.

2. Estrogens: essential hormones in female and male physiology

Estrogens are steroid hormones that bind to specific receptors, the nuclear receptor (ER) superfamily, ligand-inducible transcription factors that can bind to DNA at sites called estrogen response elements (EREs), activating different responses. Estrogens are generally recognized for stimulating cell proliferation, apoptosis, and differentiation. Although its presence has been reported in both men and women, there are three main estrogens in women, 17β-estradiol (E2) and its two metabolites (estrone and estriol), which, despite their high affinity to ER, it exerts less potent effects [1].

E2 is the most abundant; its primary source is the ovary, although it is also produced in smaller quantities in other tissues, such as the adrenal glands and adipose tissue. Estrogens are considered mainly feminizing hormones due to their effects on the female's reproductive functions; however, recently, they have also been shown to play an essential role in men, finding elevated levels of this hormone in rete testis fluid and semen where one of the most studied estrogenic effects being the negative feedback they exert on Testosterone (T) secretion [3, 4]; the production of this hormone is controlled at the hypothalamus by the Gonadotropin Releasing Hormone (GnRH), which binds to gonadotroph cells in the pituitary gland, stimulating the release of Luteinizing Hormone (LH). In addition, this hormone is released through the hypophyseal portal system, joining to their receptors in the Leydig cells located at the seminiferous tubules in the testis, where they promote the synthesis of T. So, the inhibition process that estrogen exerts on the T implies a negative feedback mechanism on the secretion of LH at the pituitary gland, causing the consequently lower secretion of T by the testicular Leydig cells [4, 5].

About their role, estrogens have been shown to participate with androgens in the maintenance of sexual behavior and testicular developmental processes. But the most critical estrogenic effect is possibly the control of the reabsorption of fluids at the epididymis, a process that causes an increase in the number of sperm in the semen and, therefore, an increase in the probability of fertilization occurring. Another vital tissue for estrogen action in men is the prostate gland, whose development and estrogenic activities influence function by controlling androgen production through ER activation. Here, the primary source of estrogen is the testis, in which Leydig cells T is transformed into E2 through an aromatization reaction catalyzed by enzymes of the cytochrome p-450 (CYP450) complex called aromatases. The E2 thus obtained is poured into the systemic circulation and reaches the target tissues, where it can be metabolized and interact in a specific way by joining the ER and intervening in direct and indirect actions [4].

3. RE: proteins responsible for the direct and indirect actions of estrogens

Two primary ER (α and β) have been identified. However, the first ER to be discovered in the 1950s was ERα, while ERβ was found almost four decades later, in 1996 [1]. These transcription factors are classified within class II of the superfamily of nuclear receptors for steroid hormones. They have a tertiary structure that includes six functionally distinct domains designated with the letters "A" to "F"; an N-terminal domain ("A/B" domains); a DNA-binding domain "C," which is the part of the receptor that binds to DNA (DNA-Binding Domain, DBD); a "D" domain, which acts as a hinge region; the "E" corresponds to the Ligand Binding Domain (LBD) and is the site of recognition and binding to the specific hormone and other related molecules; as well as a C-terminal "F" domain, whose function, although poorly understood, is believed to be involved in the conformational stability of receptors [4, 6].

It has been shown that, in the absence of a ligand, the ER remains inactive since they are bound to chaperone proteins (HSP, Heath Shock Proteins), which are responsible for maintaining the receptors in a conformation that hides their nuclear localization signal (NLS). Thus, the binding of the receptor with the corresponding hormone generates conformational changes whose consequence is the release of HSP and the appearance of nuclear localization signals; This enables the translocation of the ERs to the nucleus and the activation of the transcription of target genes by interaction with their response elements (ERE). Concerning the above, it is known that EREs are palindromic sequences with the structure 5'-GGTCAnnnTGACC-3′, where "n" represents any nucleotide. Genes with an ERE in their structure are called "estrogen-responsive" [4].

However, some genes, even lacking these sequences, can be transcribed by estrogenic signaling activated by indirect mechanisms, which will be discussed later. In this type of signaling, ERs interact with other transcription factors such as Jun, Fos, AP-1, and SP-1, causing their activation and subsequent translocation to the nucleus, where they bind with their response elements and induce gene expression. The interaction with other transcription factors is different depending on the type of ER involved, which has been proposed as an explanation for the fact that the activation of ER by identical ligands can cause other effects depending on the tissue and cellular context. These signaling mechanisms are believed to explain the actions in which ERs are involved [4, 6].

4. ERα and ERβ: receptors involved in the balance of the estrogenic effect

The discovery of estrogen receptors was made after almost four decades. This finding also required tremendous advances in the field of genetic engineering. Since the first estrogen receptor (ERα) was discovered, it was believed that this was the only one responsible for the estrogenic effects, an idea that was modified when the first transgenic mice were designed (known as "ERKO mice" < Estrogen Receptor Knock Out>) and observed that they continued to show a response to estrogenic stimulation, despite having ERα silenced. In this sense, when a patient who did not express ERα was reported in 1994, speculation began about the existence of more than one type of ER. Confirmed this assumption in 1996, the year in which the discovery of a second molecule capable of being activated by binding to its ligand (E2) and other estrogens is evidenced. Identified this new ER in rat prostate and ovary and, from then on, the first ER was named ERα, while the second is known as ERβ [1].

Both receptors are expressed on different chromosomes and show determining structural characteristics; ERα is approximately 600 amino acids long and has a molecular weight of 67KDa, while ERβ is around 500 amino acids long and weighs 55KDa, sharing about 47% amino acid sequence homology [4, 5]. Similarly, of the different domains that have been identified, the "C" domain (DBD) is the one with the most significant similarity concerning both receptors (>95%), which explains, in part, the high affinity that both receptors have for DNA [1, 6].

Specifically, although it has been reported that both receptors similarly bind to their ligands, the evidence indicates that the affinity for E2 is similar with both receptors (α and β). However, they show significant differences in binding to other natural or synthetic steroids. For example, it has been demonstrated that ERβ shows a high affinity for endogenous steroids and certain phytoestrogens (e.g., genistein). In addition to the above, it has been reported that estrogen receptors can activate different intracellular signaling pathways, which leads to their involvement in such diverse biological effects [6].

5. Ligand-dependent and ligand-independent signaling pathways activated by ERs

In general, it is known that ERs can use at least four different pathways at the cellular level. The first three pathways are ligand-dependent, and the fourth is ligand-independent. In the first "the classic route," it is known that the ligand-receptor interaction can cause ER heterodimerization, a configuration in which they can bind directly to the EREs sequences in DNA through specific binding, regulating gene expression. A second route consists of the interaction of the heterodimer after the activation of the ligand, with specific proteins, i.e., with transcription factors such as Fos/Jun, by interacting with DNA, promoting gene activation indirectly [1, 6].

In the third pathway (called the "non-genomic" type), the ligand activates the receptor to activate other transmembrane and cytoplasmic proteins, within which another receptor may be included, as well as various proteins that act as second messengers. This interaction, yet not fully elucidated, results in the opening of ion channels and the flow of ions, as well as the generation of rapid responses, including the activation of nitric oxide (NO), which has been suggested to be responsible for the physiological effects attributed to this signaling route. Finally, the fourth pathway (ligand-independent) involves crosstalk mechanisms, in which ERs are activated by other signaling pathways, such as the growth factor signaling pathway. In this case, the binding of the growth factor to its receptor promotes the activation of kinases, which, in turn, cause the phosphorylation of the ERs and their subsequent dimerization and binding to the EREs elements in the DNA [1, 6].

6. Participation of ERs in the proliferation and morphogenesis of the prostate

The different mechanisms of intracellular communication carried out by the interaction of estrogens with their ER allow them to act in several tissues, intervening in multiple functions, such as cell morphogenesis, proliferation, and differentiation. As already mentioned, it was considered that the main effects of estrogens were focused on female physiology. However, the literature had shown that the administration of environmental estrogens (xenoestrogens) from synthetic sources (e.g., bisphenol A or BPA, methoxychlor, and atrazine); pharmaceuticals (e.g., diethylstilbestrol or DES, and Ethinyl estradiol) or natural (e.g., genistein, and daidzein), caused "*in utero*" abnormalities in the urogenital tract of both males and females, with similar results being obtained when these drugs were administered at different stages of development. In addition, some studies reported other alterations in male physiology, including low semen quality, testicular neoplasia, and prostate diseases [1, 3].

In the prostate gland, both receptors have been demonstrated in epithelial cells, which control and regulate epithelial proliferation. In this sense, it has been suggested that while ERα promotes cell proliferation, ERβ limits it, so several studies have considered ERβ as a protective factor against the development of prostate diseases, while ERα could act as a promoter by being overexpressed in cancer cells; In this sense, it has been reported that the expression of ERβ is elevated in healthy tissue, while its expression decreases in damaged tissue, while the expression of ERα increases. Although the mechanisms that regulate the expression of these receptors in both conditions are unknown, it is suggested that ERβ could reduce cell proliferation by inhibiting the expression of genes involved in DNA replication and the cell cycle. In this sense, the cyclin D1 gene has been suggested as a factor that controls the transition between the G1 and S phases of the cell cycle. Another inhibition mechanism could be through direct interaction with its promoter regions or through the indirect mechanisms in which ERβ interacts with other transcription factors (AP1 and SP1, for example), stimulating or repressing the expression of their target genes. The previous is possible because some genes whose expression is controlled by ERβ activation lack EREs in their promoters. Finally, another process in which ERβ could intervene to regulate prostate cell proliferation in carcinomatous and normal cells is the expression of androgen receptors (AR), explained later in this chapter [4].

7. Estrogens and their participation in the physiology of the prostate

Development, proliferation, and function of the prostate gland depend on androgens (mainly through its primary metabolite, Dihydrotestosterone <DHT>), which require binding to its receptor (RA) for the transcription of target genes; however, it has been recognized that the prostate is sensitive to other hormones, such as steroids, which have a central role in growth, homeostasis, and in the generation of prostatic diseases [3, 7]. In this sense, it has been established that men exposed to high concentrations of xenoestrogens, whether of natural or synthetic origin, during fetal life or at some other stage of development, such exposure has a significant impact on the development of the prostate gland, so it has been shown that although estrogens can modulate epithelial morphogenesis and differentiation during prostate development, this exposure can also influence the outcome of diseases, including the benign prostatic hyperplasia (BPH) and Prostatic Cancer (PCa) [3].

As previously mentioned, these effects are regulated by the expression of ERs. Specifically in the prostate, ERα is expressed mainly in a small portion of adult prostatic stromal cells in humans, dogs, monkeys, and rodents; however, there is evidence reporting its presence in periurethral epithelial cells [6]. Furthermore, its expression is elevated in the uterine stage, declining during adolescence and adulthood, increasing again as the concentration of androgens decreases, playing a fundamental role in the changes in the prostate due to the imbalance between the T and E2 [6].

On the other hand, ERβ has been located mainly in the prostatic epithelium; Although its expression varies during fetal development, its expression increases from week 7, remaining present in low amounts throughout pregnancy, including the postnatal stage. From this period, its levels begin to rise together with the process of cytodifferentiation, reaching high levels in puberty, where it participates in prostatic epithelial differentiation [3, 7, 8].

Although it has been suggested that ER-ligand signaling is not as essential for prostate development, it is crucial for developing pathologies. As already mentioned, early treatment (especially in the prenatal stage) with estrogens or xenoestrogens produces alterations in the development and differentiation of the prostate. Furthermore, these changes occur permanently (chronic inflammation, epithelial hyperplasia, precancerous lesions, and adenomas), a process called developmental estrogenization or estrogen imprinting, which we will discuss later [7, 9]. Interestingly, it has been shown that these alterations are mediated by the signaling exerted by ERs (α and β), suggesting that the expression of these receptors can carry out antagonistic actions [2].

8. Antagonistic actions between ERβ and ERα in the generation of pathologies in the prostate

The antagonism of ERs in controlling prostate cell proliferation and differentiation has implications for the development, progression, and treatment of cancer in various tissues. There is evidence that ERα promotes while ERβ limits cell proliferation, so some studies suggest that ERβ acts as a protective factor against cancer. In contrast, ERα could be an oncogene overexpressed in carcinomatous cells [10, 11].

Evidence of this has been shown in other types of cancer, such as breast, where an absence of the chromosome that codes for ERβ has been observed, suggesting that this absence is a crucial factor in the proliferative process that accompanies tumor development. Similarly, when comparing ER expression in prostate cancer tissue, an increase in ERα expression and a decrease in ERβ expression have been observed, with a direct relationship between the reduction of this receptor and the increase in the degree of proliferation, as well as PIN (prostatic intraepithelial neoplasia) in localized prostate cancer tissues, which indicates that the absence of this receptor promotes tumor proliferation and possibly the metastatic process, having a high expression in normal epithelial cells, lower in tissue samples with BPH and very low or absent in carcinoma cells [4, 6, 12].

As we have already mentioned, this decreased expression is accompanied by an increase in the expression of ERα [10]. In this sense, it has been shown that the epithelial cells in the periurethral prostatic duct consistently express ERα in BPH tissue. Although the mechanisms accompanying these antagonistic effects are not precisely known, it is suggested that the signaling mediated by the ERβ could reduce cell proliferation by inhibiting the expression of genes involved in DNA replication and the cell cycle. Among the latter is the cyclin D1 gene, a factor that controls the transition between the G1 and S phases of the cell cycle [11].

Moreover, ERβ could inhibit the expression of these genes by directly interacting with their promoter regions and preventing their transcription or through indirect mechanisms in which ERβ interacts with other transcription factors (AP1 and SP1, for example) and causes stimulation/repression of their target genes. The last is possible because some genes whose expression is controlled by ERβ activation lack EREs in their promoter regions. More aggressive PCa have recently been found to be associated with the presence of an acquired chromosomal mutation. In this translocation, the promoter region of the Serine-Protease Transmembrane-2 (SPT2) gene is fused with the coding region of some transcription factors of the Erythroblast Transformation-specific (EST) family. Chromosomes with this fusion (SPT2-EST) are susceptible to the action of ERs. Specifically, activation of ERα has been shown to promote an increase in SPT2-EST gene expression, while activation of ERβ promotes a decrease [4, 10].

Another process in which ERβ intervenes to regulate proliferation in both carcinomatous and normal cells is the expression of ARs; cell proliferation depends mainly on androgens, which require their receptor to exert their effects by activating the target gene transcription [4]. Some studies show that one of the functions of ERβ is to maintain AR expression at adequate levels to prevent excessive cell proliferation. For example, in βERKO mice, there are higher amounts of AR than in healthy mice, as well as an increase in cell proliferation. Similarly, treating human prostate cancer cell cultures with ERβ-specific ligands results in decreased AR expression [10]. In this context, it is possible to promote an increase in the expression of AR by administering specific agonists of the ERα, which constitutes another evidence of the antagonism between these two receptors in the development of pathologies. In the same way, because the ERβ can act in different stages of development, intervening in processes such as cell differentiation, antiproliferative, anti-inflammatory, and antioxidant actions is a good candidate for developing therapeutic strategies in tumor processes [4, 6].

9. Environment, epigenetics, and estrogens; how this influences prostate development

The morphogenic development of the prostate gland depends on the participation of multiple components, including hormones and a wide variety of factors, among which the Homeobox genes stand out, which includes the *Hox* and *NK* family genes. Within these genes are *Hox*b-13 and *Nk*x3, whose expression is essential for prostatic epithelial differentiation. Interestingly, it has been shown that steroid hormones can regulate these genes. Specifically, exposure to the xenoestrogen DES has resulted in the downregulation of *Hox*b-13 gene expression and suppression or downregulation of the *Nk*x3 gene. The last is correlated with a loss of prostatic epithelial differentiation and suppression of secretory genes, triggering alterations related to the estrogenization process that occurs in later stages [7, 8]. In addition to the above, embryogenesis requires the presence of proteins expressed at the level of epithelial-mesenchymal tissue cells, as well as the signals exchanged by these tissues. *Bmps* proteins act as inhibitors of proliferation during development; specifically, the *Bmp*4 mRNA is expressed in mesenchymal tissue during embryonic development, declining in the first days of postnatal development. Interestingly, exposure to estrogens in the prenatal period causes *Bmp*4 protein levels to remain elevated for a prolonged period after birth (day 15), the direct consequence of which is associated with prolonged suppression of prostate tissue, which contributes to the development of a hypomorphic phenotype [7, 8].

Similarly, other morphogenic developmental regulators are the *Shh* and *Fgf* 10 proteins. On the one hand, *Shh* is a secretory glycoprotein produced by epithelial cells in developing tissues. At the same time, the *Fgf* 10 factor has a central role in stimulating prostatic budding, ductal outgrowth, and branching morphogenesis. As expected, the presence of prenatal estrogens produces an inhibition of both molecules (*Shh* and *Fgf* 10), which results in an inhibition of branching morphogenesis [7, 8].

Furthermore, in the adult stages of development, other epigenetic factors influence the appearance of prostate alterations; In this sense, reports are indicating that in cancerous tissues, the ERα gene is methylated, which leads to a silencing of this gene, triggering different types of alterations, ranging from the development of BPH to varying degrees of tumor lesions. Therefore, it has been suggested that the risk of PCa is highly associated with polymorphisms in the ERα gene, particularly in populations such as the Japanese and African, which are more susceptible to estrogenic actions [6]. Similarly, it has been reported that the absence/decrease in the expression of the β receptor that occurs as the cancerous proliferative process progresses could be associated with a hypermethylation process, which has been shown to appear in the CpG islands of the promoter region of the gene that codes for this receptor, causing the silencing of this gene and its transcription. This effect is reversed with demethylase treatment, an epigenetic mark involved in the progression of prostate tumors [13].

10. Conclusions

The prostate is a gland responsive to estrogens, which are involved in the normal development of this organ (morphogenesis) and the development of pathologies. These actions are regulated at different stages of development by the presence/absence of ERs (α and β) and the pathways activated by them, participating both in normal tissue development and in the generation of pathologies.

Acknowledgements

The author thanks to the National Council of Science and Technology (CONACyT) for the supporting founding.

Conflict of interest

The author declares no conflict of interest.

Acronyms and abbreviations

ER α/β	estrogen receptors α and β
BPH	benign prostatic hyperplasia
PCa	prostate cancer
E2	17β-estradiol
EREs	estrogen response elements
T	testosterone
GnRH	gonadotropin releasing hormone
LH	luteinizing hormone
CYP450	cytochrome p-450
DBD	DNA-binding domain
LBD	ligand binding domain
NLS	nuclear localization signal
HSP	heath shock proteins
ERKO	estrogen receptor knockout mouse
NO	nitric oxide
BPA	bisphenol A
DES	diethylstilbestrol
AR	androgen receptor
DHT	dihydrotestosterone
PIN	prostatic intraepithelial neoplasia
SPT2	serine-protease transmembrane-2 gen
EST	erythroblast transformation-specific factor

Author details

José Locia Espinoza and Luz Irene Pascual Mathey*
Faculty of Biology of Pharmaceutical Chemistry, Veracruzana University,
Xalapa, Veracruz, Mexico

*Address all correspondence to: lupascual@uv.mx

References

[1] Heldring N, Pike A, Andersson S, Matthews J, Cheng G, Hartman J, et al. Estrogen receptors: How do they signal and what are their targets. Physiological Reviews. 2007;**87**(3):905-931. DOI: 10.1152/physrev.00026.2006

[2] Hess RA, Cooke PS. Estrogen in the male: A historical perspective. Biology of Reproduction. 2018;**99**(1):27-44. DOI: 10.1093/biolre/ioy043

[3] Cooke PS, Nanjappa MK, Ko C, Prins GS, Hess RA. Estrogens in male physiology. Physiological Reviews. 2017;**97**(3):995-1043. DOI: 10.1152/physrev.00018.2016

[4] Espinoza JL, Aguilar MEH, Abreu GEA, Durán FR, Denes JM, Ávila GAC, et al. El papel de los estrógenos y sus receptores en la prevención y promoción de enfermedades proliferativas de la glándula prostática. Eneurobiología: Revista electrónica. 2013;**4**(8):1-23. Available from: http://www.uv.mx/eneurobiologia/vols/2013/8/Locia/HTML.html

[5] Sharpe R. The roles of oestrogen in the male. Trends in Endocrinology & Metabolism. 1998;**9**(9):371-377, ISSN 1043-2760. DOI: 10.1016/S1043-2760(98)00089-7

[6] Prins GS, Korach KS. The role of estrogens and estrogen receptors in normal prostate growth and disease. Steroids. 2008;**73**(3):233-244. DOI: 10.1016/j.steroids.2007.10.013

[7] Prins GS, Huang L, Birch L, Pu Y. The role of estrogens in normal and abnormal development of the prostate gland. Annals of the New York Academy of Sciences. 2006;**1089**:1-13. DOI: 10.1196/annals.1386.009

[8] Prins GS, Birch L, Tang WY, Ho SM. Developmental estrogen exposures predispose to prostate carcinogenesis with aging. Reproductive Toxicology (Elmsford, N.Y.). 2007;**23**(3):374-382. DOI: 10.1016/j.reprotox.2006.10.001

[9] Prins GS, Birch L. Neonatal estrogen exposure up-regulates estrogen receptor expression in the developing and adult rat prostate lobes. Endocrinology. 1997;**138**(5):1801-1809. DOI: 10.1210/endo.138.5.5106

[10] Bonkhoff H, Berges R. The evolving role of oestrogens and their receptors in the development and progression of prostate cancer. European Urology. 2009;**55**(3):533-542. DOI: 10.1016/j.eururo.2008.10.035

[11] Bardin A, Boulle N, Lazennec G, Vignon F, Pujol P. Loss of ERbeta expression as a common step in estrogen-dependent tumor progression. Endocrine-Related Cancer. 2004;**11**(3):537-551. DOI: 10.1677/erc.1.00800

[12] Horvath LG, Henshall SM, Lee CS, Head DR, Quinn DI, Makela S, et al. Frequent loss of estrogen receptor-beta expression in prostate cancer. Cancer Research. 2001;**61**(14):5331-5335

[13] Zhu X, Leav I, Leung YK, Wu M, Liu Q, Gao Y, et al. Dynamic regulation of estrogen receptor-beta expression by DNA methylation during prostate cancer development and metastasis. The American Journal of Pathology. 2004;**164**(6):2003-2012. DOI: 10.1016/s0002-9440(10)63760-1